Sacred Scents in Early Christianity and Islam

Studies in Body and Religion

Series Editors:

Richard M. Carp, Saint Mary's College of California
and
Julia Huang Lemmon, National Tsing Hua University, Taiwan

Studies in Body and Religion publishes contemporary research and theory that addresses body as a fundamental category of analysis in the study of religion. Embodied humans conceive of, study, transmit, receive, and practice religion, with and through their bodies and bodily capacities. Volumes in this series will include diverse examples and perspectives on the roles and understandings of body in religion, as well as the influence and importance of religion for body. They will also move conversation on body and religion forward by problematizing "body," which, like "religion," is a contested concept. We do not know exactly what religion is, nor do we know exactly what body is, either; much less do we understand their mutual interpenetrations. This series aims to address this by bringing multiple understandings of body into an arena of conversation.

Titles in the Series:

Early Daoist Dietary Practices: Examining Ways to Health and Longevity, by Shawn Arthur
Dancing Culture Religion, by Sam Gill
Risky Marriage: HIV and Intimate Relationships in Tanzania, by Melissa Browning
Dancing Bodies of Devotion: Fluid Gestures in Bharata Natyam, by Katherine C. Zubko
Sacred Scents in Early Islam and Christianity, by Mary Thurlkill

Sacred Scents in Early Christianity and Islam

Mary Thurlkill

LEXINGTON BOOKS
Lanham • Boulder • New York • London

Published by Lexington Books
An imprint of The Rowman & Littlefield Publishing Group, Inc.
4501 Forbes Boulevard, Suite 200, Lanham, Maryland 20706
www.rowman.com

Unit A, Whitacre Mews, 26-34 Stannary Street, London SE11 4AB

British Library Cataloguing in Publication Information Available

Library of Congress Cataloging-in-Publication Data

Names: Thurlkill, Mary F., 1969- author.
Title: Sacred scents in early Christianity and Islam / by Mary Thurlkill.
Description: Lanham : Lexington Books, 2016. | Series: Studies in body and religion | Includes bibliographical references and index.
Identifiers: LCCN 2016015833 (print) | LCCN 2016016268 (ebook) | ISBN 9780739174524 (cloth : alk. paper) | ISBN 9780739174531 (Electronic)
Subjects: LCSH: Human body--Religious aspects--Christianity. | Senses and sensation--Religious aspects--Christianity. | Smell--Religious aspects. | Odors. | Smell--Religious aspects--Christianity. | Smell--Religious aspects--Islam.
Classification: LCC BL65.B63 T48 2016 (print) | LCC BL65.B63 (ebook) | DDC 203--dc23
LC record available at https://lccn.loc.gov/2016015833

∞™ The paper used in this publication meets the minimum requirements of American National Standard for Information Sciences Permanence of Paper for Printed Library Materials, ANSI/NISO Z39.48-1992.

Printed in the United States of America

For Cody Thurlkill

Table of Contents

Preliminary Notes

TRANSLATIONS AND TRANSLITERATIONS

Located at the University of Mississippi, with a small library collection devoted to medieval primary sources, I have taken full advantage of on-line resources and translations. Most significantly, instead of following the standard A. J. Wensick's *Concordance et Indices de la Tradition Musulmane,* I use www.sunnah.com. Even though the site incorporates various translators, it also includes the Arabic for cross-checking. Thus, in citing individual hadith, I include the collection (e.g., *Bukhari* or *Muslim*) followed by the in-book reference noted at www.sunnah.com.

Hadith collections referenced at www.sunnah.com include:

Sahih al-Bukhari
Sahih Muslim
Sunan an-Nasa'i
Sunan Abi Dawud
Jami` at-Tirmidhi
Sunan Ibn Majah
Muwatta Malik

For early Christian sources, I generally provide references to the on-line *Early Church Fathers* translation series available at Christian Classics Ethereal Library (www.ccel.org). These translations are not always the most elegant, but they are widely accessible for interested readers.

Because this book is aimed at readers for both early Christianity and Islam, I also have standardized as many Arabic transliterations as possible. I do not include the macron or underdot in the body of the text; I only include diacriticals in the notes, following the transliteration guide adhered to by the *International Journal of Middle East Studies*.

DATES

I include the standard Gregorian dating system throughout the work. All Islamic dates (AH) are converted to common era (CE); although, I occasionally include both for clarity.

Abbreviations

AS	*Acta Sanctorum.* Brussels: Impression Anastaltique Culture et Civilisation, 1970.
CCSL	*Corpus Christianorum. Series Latina.* Turnholt: Brepolis Editores Pontificii, 1956.
CSEL	*Corpus Scriptorum Ecclesiasticorum Latinorum.* New York and London: Johnson Reprint, 1963.
EI^1, EI^2	*Encyclopedia of Islam*, 1st and 2nd eds. Leiden: E. J. Brill.
IJMES	*International Journal of Middle East Studies.* London: Cambridge University Press.
M, T, B, J	*Mishnah, Tofseta, Babylonian Talmud, Jerusalem Talmud* (respectively)
MGH	*Monumenta Germaniae Historica.* Societas Aperiendis Fontibus Rerum Germanicarum.
SRM	*Scriptores Rerum Merovingicarum.*
PG	J. P. Migne, ed., *Patrologia Graeca,* Paris, 1886.
PL	J. P. Migne, ed., *Patrologia Latina,* Paris, 1886.
SC	*Sources Chrétiennes.* Paris: Editions du Cerf, 1943–.

Introduction

A few years ago, some friends asked that I check on the family pets while they traveled during Christmas break. I arrived late on the first evening, confused by the layout of their new home. I entered the front door, unable to locate a light switch, so I wandered across the living area in complete darkness, basically by touch—with no sound and no light. After I had crossed about half the room, in complete darkness, I encountered their cedar Christmas tree's incredibly strong scent. I had not smelled a "real" Christmas tree in years, and it immediately brought me back to my childhood, complete with fond memories of holiday traditions. The scent prompted an unusually emotional response, and I stood rather stunned and motionless for several minutes.

Not long after this experience, I sat in my office, reading through some medieval Shi`ite Islamic hadith (or sayings of the Prophet) that related to Muhammad's love for his daughter, Fatima. The Prophet described the happiness she brought him and noted that when he walked past her, "he smelled the smells of Paradise."[1] This saying conveys the distinctively Shi`ite theology that elevates `Ali (the Prophet's cousin and son-in-law), Fatima, and their children as Imams, or perfect human beings that may intercede between the believer and Allah. According to Shi`ite tradition, Muhammad "conceived" Fatima by eating a fruit proffered by Gabriel during his *mir'aj*, or ascent into Paradise. After he ate the fruit (usually identified as a date), he returned to earth, had sexual intercourse with his wife Khadija, and deposited Fatima's "radiance" into her womb. Fatima's presence later delighted Muhammad as he remembered his heavenly journey and encounter with God.

Reading the hadith reminded me of my own recent "tree" event, and thus commenced this project on scent. Muhammad experienced a joyful memory, viscerally based—in the body and its senses—that effectively communicated across time and space. Understanding the body and its senses helps "make sense" of various cultures, rituals, geographies, and theologies. Indeed, Sensory Anthropology informs more than ethnographic fieldwork—how researchers should employ all their senses for more nuanced observations—and provides a valuable tool in approaching historical and religious texts as well.

According to David Howes and Constance Classen, two pioneers in Sensory Anthropology at Concordia University, responsible cultural and social theory requires scholars to consider seriously the body as cultural expression and understand its sensations as more than just personal physiological experience.[2] Contemporary Western readers too often correlate the body with "individual" and "private"; for example, my smelling a cedar tree in my friend's dark house triggered a series of very personal memories. That may be true, but the matter is far from done. That cedar tree's meaning is deeply embedded within historical constructions that relate much about social status (who can afford the luxury?), gender (who attends to domestic rituals?), and religion (Christmas celebrations in the American South).

Sensory Anthropology recognizes that the body's sensorium links with constantly shifting cultural meanings emergent within distinctive time and space/geography; yet it contends that sensual encounters help create those cultural meanings as well. One early academic study of scent, Alain Corbin's *Foul and Fragrant*, for example, inspects the complex connections between social power and odor in eighteenth- and nineteenth-century France.[3] Corbin tracks developments in healthcare, urban planning, and bourgeois class identity through scientific accounts of the "stinking" sick and elite repugnance toward the "dirty" poor. This groundbreaking work allows for a subtle portrait of early modern French culture's shifting power relations instead of a simplified survey of economic and social crises. Focusing on scent also compels academic discourse to move beyond an epistemology rooted in the Western privileging of only sight/vision/text for more nuanced cultural readings.

Recent works by Religious Studies scholars and historians have built upon such studies. Susan Ashbrook Harvey's masterful *Scenting Salvation: Ancient Christianity and the Olfactory Imagination* traces olfactory experiences in late antique and early medieval Christianity in the Greek and Syriac East.[4] She describes how an emerging Christian community defined and even reimagined itself through the sensorium. Her most extraordinary example presents the Syrian holy man, joined in Christ's suffering through his voluntary deprivations, imagined as the "sweet-smelling sacrifice" to God despite his body's stench and decay.[5] Not surprisingly, the Stylite mirrors Christ's own physical suffering, providing a theological graphic for potential converts. Yet, by upending the audience's expectations of sweet smell,

Christian authors also critiqued ordinary, worldly sensations and promised the mysterious (and spiritual) delights awarded to Christian bodies.

By focusing on olfaction, Harvey surveys ancient uses of scent in both sacred and profane contexts and explains how Christians conceived of incense, sacrifice, and embodied practice (especially asceticism).[6] Mark Bradley and Shane Butler recently added to the discussion with their *The Senses in Antiquity* series (Routledge) that centers on Greco-Roman culture more generally. One volume addresses the entire synaesthesia and the other three focus on sight, smell, and touch independently. The volume devoted to odor includes two particularly important articles demonstrating how rabbinic Jews and post-Constantinian Christians employed scent in defining their relationship *with* the Divine and *against* their "pagan" counterparts.[7]

Yael Avrahami also tackles the entire sensorium in *The Senses of Scripture: Sensory Perception in the Hebrew Bible.* [8] Avrahami ably explores Biblical epistemology, the abundant use of sensual language in Biblical tradition, and the primary correlation between sight/hearing and Divine "knowing." Avrahami suggests, for example, that Biblical tradition motivated the West's preoccupation with sight/vision/text just as deeply as Plato and other Greco-Roman philosophers. Deborah Green's *The Aroma of Righteousness* returns to scent as a primary conveyer of meaning in the Hebrew Bible and rabbinic discourse.[9] She situates fragrant symbolism and its related material culture within the rabbinic world and offers a feminist critique of their implications. Green's analysis follows the rabbis intimate connection with the Divine forged through scented ritual as well as their appropriation of Biblical sensual imagery.

These recent works recognize scent as an important lens for historical and cultural inquiry, successfully challenging the traditional Western privileging of epistemological hierarchies that first rank mind/reason above matter/emotion and, second, classify scent among the "lower" senses.[10] Plato initially understood the body's senses as the source of knowledge that proved both confusing and even deceptive. The mind, on the other hand, could attain the realm of *ideas*, or eternal Truths. Aristotle inherited this ranking although he focused more on the sensorium, ordering them as "human" (including sight, hearing, and smell) and "animal" (taste and touch).[11] Most modern theories of philosophy and psychology further demote scent as animalistic, uncivilized, and boorish, best left to humanity's evolutionary past where it linked with sexuality and reproduction.[12] Current scholarship thus reverses these trends, focusing on embodiment and sensual religious practice.

In this present work, I aim to contribute to this lively conversation on the sensorium and sacred scents in particular. Following Green, I examine quotidian sensory encounters in civic and domestic space, cooking manuals, and medical treaties by drawing on history, anthropology, and archaeology/material culture (Part I). I incorporate a wide array of sources beginning with the

Greco-Roman and pre-Islamic Arabian cultural milieus from which early Christian and Muslim identities spring. The ancient Mediterranean world enjoyed many of the same aromatics because of the Silk Road networks; however, early Christians and Muslims interpreted those scents in distinctive ways. Pliny the Elder's (d. 79 CE) *Natural History*, for example, provides basic clues about spices and perfumes available through international trade. Part I—"Sensory Worlds"—references Greco-Roman, Jewish, and pre-Islamic Arabian sources in describing how ancients applied these scents in civic and domestic space. Herein, we will survey the aromatics available to early Christians and Muslims as well as explore their multiple meanings. When hosts provided fragrant water for guests to bathe their hands and feet before sharing a meal, scent encoded a hierarchical relationship appreciated by both parties. Gender ambiguity occurred when men smelled like women; and, too much fragrance exposed women as wonton whores. Scent effectively cued every social relationship and gender expectation relevant for ancient bodies.

In Part II—"Sacred Scents"—I concentrate more on religious meanings and sacred texts, including liturgical manuals, theologies, and hagiographies (or, writings about saints). Early Christians eagerly employed aromatics in their rituals only after Emperor Constantine legitimized the faith in 313 CE. Initially Christian leaders considered sacred scents and perfumes as too "pagan"; yet, by the end of the fourth century, incense shrouded the altar and spiced oils anointed initiates. The Prophet Muhammad also equated his pious followers with sweet smells even though he did not require scent as part of the daily prayer routine. Both nascent religious groupings generally employed scent as an identity marker, and a "rhetoric of disgust" against the heretical other.

This work concludes with Part III—"Scents of Paradise"—an examination of a sacred Garden/Heaven, the reward for religious piety and faithfulness. For both Christians and Muslims, Paradise represented perfect unity enjoyed between humanity and God that might only be recovered at the end times. Paradise, imagined as a sublime Garden, instantiates the fragrant archetypal existence, eternal and pure, abandoned in willful disobedience by Adam and Eve, resulting in death's stench and decay. While contemplating the divine return allows Christians and Muslims to imagine a joyous reunion, the embodied rituals and renderings of Paradise initiate a sensual longing, a desire to smell, touch, taste, hear, and see God again.

While inspired by Green and Harvey's works on smell in terms of methodology, this work uniquely moves comparatively in both its geography and theology. I focus on the formation and evolution of early Christian identities beginning in the Greco-Roman world through, mainly, the Merovingian period (c. 800 CE). In Islam, I begin with pre-Islamic poetry and work through the classical world (c. 1250 CE, with the Umayyad and `Abbasid dynasties), although I include later sources when they offer insight into earlier practice and thought. Such a comparative approach does not suggest linear dissemination, transmission, or "cultural borrowing." Instead, I examine both similarities and distinctions relating to embodiment and the sensorium among two separate, emerging religious systems.

Religion scholar Jonathan Z. Smith has argued that comparative studies offers another theoretical tool to learn something "new" by introducing a third category of analysis. Instead of presenting a facile list of similarities and differences, comparative religion proposes a third term to "make meaning." In this study, I examine scent in early Christianity and early Islam and introduce "olfaction and transition" as a comparative category.[13] This complements the selected time frame as well; Christian and Islamic identities themselves are "*in the making*" in late antiquity and the early Middle Ages. Scent marks movement and transition amid different spaces, times, and concepts. For example, sweet smells signal the distinction between sacred and profane space and time for Christians and Muslims: incense sanctifies the altar and sweet perfumes waft from garments at Friday sermon. By better understanding the embodied, sensory experiences presented in ritual, text, and material culture, we uncover a web of meaning indicating emerging religious identities, ethical guidelines, and gender hierarchies.

Scent provides a particularly powerful marker of transition, liminality and (re)-imagination because of its ephemeral nature.[14] Unlike other sensory encounters, scent proves more difficult to identify, locate at its source, and control. It penetrates bodies and boundaries without permission and resists limits. It refuses easy classification and instead "smells like" something else, an infinite regress of precise meaning. Because of the physiology of scent, it links directly to the brain's limbic system after molecules enter the nose. The limbic system, the depository of emotions and memory, stores odor within its nexus. Thus scent experiences can instantly and dramatically transport us to the past, reimagining us across time and space.[15] Both Christian and Islamic traditions, for example, interpret a corpse's sweet smell as proof of sanctity—it recalls at once the primordial Garden from which humanity descends as well as the eschatological Paradise that awaits the pure. Both traditions place the saint in the presence of God even though the body remains on earth. Scent collapses time and space in sensuous imagination as eternity shrinks to the present in the company of the saints.

Finally, because odor expands outside its original source, it effectively unites its audience in a sensory experience, linking the individual body to the body social. While aromatics might prompt an array of intimately personal emotions and memories, they also promote group solidarity and a sense of inclusiveness. Here, we can distinguish scents' social meaning more clearly as they cue class status, gender normative and queer behaviors, morality, and relatedly, orthodoxy. Both Christian and Islamic traditions, for example, style pious men and women as fragrant offerings to God; however, horrid stench often betray those who stray outside evolving theologies and orthodoxies.[16] Christian and Muslims thus employed scent to mark their own emerging religious identities as well as the heretical "other" that threatened social purity and orthodoxy. These divisions appear most poignantly in a "rhetoric of disgust" which imagines the "other" as the source of vile filth and pollution to the body social. As we shall see, surveys of hell's wretched tortures harmonize with heaven's opulent beauties to demonstrate the fate awaiting believers and unbelievers alike. Scent cued not only relevant social and cultural distinctions in early Christianity and Islam but eternity itself.

NOTES

1. Al-Majlisī, *Biḥār al-anwār* (Tehran-Qum, 1376–1392/1956–1975), v. 43.1.4.
2. Their joint works include the recent *Ways of Sensing: Understanding the Senses in Society* (London: Routledge, 2013). Another important introductory work is *The Varieties of Sensory Experience: A Sourcebook in the Anthropology of the Senses* (Toronto: University of Toronto Press, 1991), ed. David Howes. Paul Stoller's *The Taste of Ethnographic Things* (Philadelphia: University of Pennsylvania Press, 1989) addresses the privilege of sight over the full sensorium in this early work.
3. Alain Corbin, *The Foul and the Fragrant: Odor and the French Social Imagination* (Cambridge, MA: Harvard University Press, 1986). Corbin focuses particularly on scent's literary relation to social class.
4. Susan Ashbrook Harvey, *Scenting Salvation: Ancient Christianity and the Olfactory Imagination* (Berkeley: University of California Press, 2006).
5. "On Holy Stench: When the Odor of Sanctity Sickens," *Studia Patristica: Papers Presented to the International Conference on Patristic Studies*, v. 35 (1998): 90–101.
6. Some other important works dealing with the entire sensorium in medieval Christianity include Rachel Fulton, "'Taste and see that the Lord is sweet' (Ps. 33:9): The Flavor of God in the Monastic West," *Journal of Religion* 86.2 (2006): 169–204; Rosemary Drage Hale, "'Taste and See, for God Is Sweet': Sensory Perception and Memory in Medieval Christian Mystical Experience," in *Vox Mystica: Essays on Medieval Mysticism in Honor of Professor Valerie M. Lagorio*, ed. Anne Clark Bartlett (Rochester: D. S. Brewer, 1995), 3–14; Boyd Taylor Coolman, *Knowing God by Experience: The Spiritual Senses in the Theology of William of Auxerre* (Washington, DC: Catholic University of America Press, 2004); and Gordon Rudy, *Mystical Language of the Senses in the Later Middle Ages* (New York: Routledge, 2002).
7. Deborah Green's "Fragrance in the Rabbinic World" and Jerry Toner's "Smell and Christianity" in *Smell and the Ancient Senses*, ed. Mark Bradley (New York: Routledge, 2015).
8. Yael Avrahami, *The Senses of Scripture: Sensory Perception in the Hebrew Bible* (New York: T&T Clark, 2012).
9. Deborah Green, *The Aroma of Righteousness: Scent and Seduction in Rabbinic Life and Literature* (University Park: Pennsylvania State University Press, 2011).
10. Paul Stoller, a pioneer in "sensory anthropology," challenges Western "superiority of sight" in his groundbreaking *Taste of Ethnographic Things* (Philadelphia: University of Pennsylvania Press, 1989).
11. Aristotle addresses the senses and their ordering, particularly, in *Nicomachean Ethics* and *Eudemian Ethics*. Also, see an overview of ancient philosophical views of the senses in Han Baltussen, "Ancient Philosophers on the Sense of Smell," *Smell and the Ancient Senses*, ed. Mark Bradley (New York: Routledge, 2015), 30–45.
12. Notably Hegel's *Aesthetics*, 728–37; Annick LeGuérer pulled together Freud's references to smell and sexuality in his "Le declin de l'olfactif: mythe ou réalité?," *Anthropologie et Sociétés* 14 (2): 25–46. Also, see Hans J. Rindisbacher, *The Smell of Books: A Cultural-*

Historical Study of Olfactory Perception in Literature (Ann Arbor: University of Michigan Press, 1992), for a concise discussion of sensory hierarchy in Western philosophy and psychology. Another important overview is Anthony Synnott's "The Senses: The Puzzle and the Balance," in *The Body Social: Symbolism, Self, and Society* (New York: Routledge, 1993), chapter 5 (128–55).

13. I am inspired here by Howes, "Olfaction and Transition," in *The Varieties of Sensory Experience*, 128–47.

14. I rely here, of course, on Victor Turner's ritual theory as discussed in *The Ritual Process: Structure and Anti-Structure* (Chicago: Aldine Transactions, 1969).

15. Trugg Engen offers an important and accessible discussion of human reaction to odor in *Odor Sensation and Memory* (New York: Praeger, 1991).

16. For important examples of this approach to embodied religion, see James Aho who builds upon Mary Douglas' theory of "dirt out of place" and social integrity; see *The Orifice as Sacrificial Site: Culture, Organization and the Body* (Dulles, VA: AldineTransaction, 2002).

Part I

Sensory Worlds

Before we can fully understand scent's significance in early Christianity and Islam, for both the human body and body social, we must first situate it within its cultural and historical framework. Ancient Mediterranean Jews, Christians, and Muslims enjoyed mostly the same aromatics—and assigned them similar meaning—because of profitable trade. The spice route, beginning in the first millennium BCE, successfully united Arabia with Africa and Europe. Trade followed two basic paths: first, caravans conveyed spices and incense grown along the southern Arabian coast and procured in India and Southeast Asia to Gaza, where merchants shipped them to the Mediterranean;[1] second, ships from the South Arabian port city Qana' ferried loads of aromatics to Roman Egypt where it proliferated throughout the Empire. Then, in 17 CE, Greek sailor Hippalus discovered how to harness monsoon winds to directly connect the Eastern Mediterranean with India, thus linking the Roman Empire with the East more effectively. According to geographer Strabo (d. c. 24 CE), approximately 120 Roman ships made the year-long round trip journey to India annually.[2]

Mediterranean and Arabian cultures had always treasured sweet smells, even before the spice routes' improved efficiency; until then, ancient communities simply found their scent closer to home. The Levant, for example, boasted aromatic balsam, saffron, and henna.[3] Mesopotamia produced oils scented with myrtle, cypress, opopanax, and fragrant reed. Egypt relied upon oils pressed from colocynth, horseradish, sesame, and olives.[4] The spice routes allowed for more variety, including frankincense and myrrh (mostly from southern Arabia), and cinnamon, pepper, cardamom, and sandlewood from India and Southeast Asia.

The spice trade, at its zenith between the second century BCE and the second century CE, provided not only an impressive variety of scents but also moderate accessibility. While wealthy Romans and Arabs secured the finest aromatics available, textual and archeological evidence suggest that most social groups utilized sweet scents for daily life and sacred purpose. Poorer classes made due with fewer choices and lower qualities of merchandise. Frankincense, for example, originates from more than twenty-five different species of trees.[5] Wealthy customers could afford the finest quality, with the cleanest "burn" and purest scent, while others endured smokier versions. According to Greek botanist Theophrastus (d. 386 BCE), oils provided the best bases for perfumes, with ben oil (garnered from nuts found on desert bushes) in the most valuable and green olive oil in the most common perfumes.[6] Precious spices used for cooking, medicine, and sacred ritual also signaled social class: pepper, cinnamon, and cassia (among others) from India and Southeast Asia probably flavored the lives of only the elite.[7]

Of course, these class distinctions assume that customers always got what they paid for: natural philosopher Pliny the Elder (d. 79 CE) notes that merchants often lied about their products and adulterated them to increase their yield and their profits. Frankincense, for example, oozes from the sliced bark of several different species of trees and puddles along the tree's trunk or base. Workers cultivated the purest—and most costly—frankincense from certain species that they sliced in summer and harvested in the autumn. According to some of Pliny's sources, workmen only came in contact with these premier crops when in a state of purity, meaning they had no immediate contact with women or the dead.[8] With such criteria, only the most refined customers would be able to evaluate the product's true quality. Pliny even includes some of the more well-known recipes for adulteration: Indian pepper mixed with cheap Alexandrian mustard, white resin with frankincense and, of course, combining cheaper qualities of the same product, such as myrrh.[9]

Regardless of each social class' distinct commodities, Romans and Arabs generally put spices, scents, and aromatics to similar use. Too many historians classify fragrance and spice as luxury items, nice to have around but hardly necessary, and available only to the wealthy. In Part I, we will examine Roman and Arab cultures just before and during Christianity and Islam's rise and proliferation. We will catalog available aromatics and spices and define how they functioned among distinct social classes by focusing on three different categories, divided into three separate chapters.[10] Chapter 1 surveys scents' routine civic and domestic use in personal hygiene, homes, public gatherings, and royal courts in both the Roman West and Arab Middle East. Romans, for example, applied aromatics at baths, public games, and sporting events. Arabians sprinkled crushed musk and incense in their pockets and around their cushions and carpets. In both realms, sweet smells enhanced everyday life by providing pleasurable fragrance; protection against sickness and evil; and improved sexual gratification. At Roman home altars, worshippers offered incense to their gods, and both groups included fragrance at important life rituals such as marriage and death. The two political cultures associated fragrance with peace and included it in gift-giving while forging new alliances and pacts. In both time and space, scent often signaled *transition*—passage from mundane to sacred, public to private, war to peace.

In chapter 2 we will look at scents and spices' culinary uses in both flavoring and aromatizing food. Even though the kitchen can be included within "domestic use," cooking manuals in the ancient and medieval worlds resembled treatises on society and health issues more than their modern counterparts. Ancients—outside the eremitic ideal—seldom ate alone; food consumption occurred in public, among family, friends, and powerful patrons. Food rituals pervaded Christian and Islamic cultures and defined social status as well as religious orthodoxy. Romans and Arabs alike associated certain culinary categories with conquerors versus the conquered and believers versus heretics. This becomes evident in the incredibly complex recipes located in cookery books.

Finally, chapter 3 explores how the ancient and medieval Mediterranean world associated sweet smells with healing; indeed, most cookbooks provided home remedies and popular curatives, so this section relates closely with the previous one. Ancient Romans and Arabs relied upon their natural world for nutritional, transformative healing. Observers often intuited from nature itself their surrounding world's curative properties. Balsams such as frankincense and myrrh, for example, exude pleasant scent after they seep from damaged tree bark. The resinous substance also resists general patterns of putrefaction. After observing these qualities, the ancients applied resins to heal wounds and preserve life.[11] Likewise, healers reckoned that balsams provided antidotes for snake bites because vipers enjoyed their sweet juice and shade; thus, they utilized the fragrant plants in alleviating toxic bites.[12] Healing scents brought more than restoration; they transformed the body—a kind of alchemy—that made a broken form whole and perhaps even immortal. Here again—as with other routine uses of aroma—scent signaled change, shift, and renovation.

NOTES

1. As Richard Bulliet points out, this overland route offered traders more economic incentive after Arabs adapted a new harness for camels. See *The Camel and the Wheel* (Cambridge, MA: Harvard University Press, 1975).
2. Discussed in Jack Turner, *History of a Temptation* (New York: Alfred A. Knopf, 1994), 59.

3. Egyptian Maṭariyya (near Heliopolis) was particularly famous for its fine balsam. According to one Christian tradition, as the Holy Family sojourned in Egypt after they fled Jerusalem, Christ took Joseph's staff, broke it into pieces, and planted the pieces in the soil. Water then miraculously erupted and sections of the staff sprouted into fragrant leaves. Joseph's staff, constructed in Jericho, provided the narrative link between the sweet Levantine balsams and their Egyptian counterparts. See Milwright's discussion, 204–5.

4. See Jean-Pierre Burn's discussion in "The Production of Perfumes in Antiquity: The Cases of Delos and Paestum," *American Journal of Archaeology* 104.2 (April 2000), 278–80.

5. David Peacock and David Williams, *Food for the Gods: New Light on the Ancient Incense Trade* (Oxford: Oxbow Books, 2007), 21. This variety is evident in Pliny's *Natural History,* 12.31, as he discusses the confusion over the exact characteristics of frankincense trees. According to him, no one could agree upon color, leaf shape, etc.

6. Theophrastus, *De odoribus*, §§ 14–16, as discussed in Burn, 281–82. Also, Horace bemoans the fact that wealthy Romans transformed small farms and ancient olive groves with large estates, complete with gardens filled with "myrtle, violets and other sweet-smelling flowers." *Odes,* ii.15.

7. Pliny describes one of the more innovative techniques in harvesting spices with cassia. According to him, workmen trimmed the ends of branches and sewed them within fresh skins of cattle. As the skin putrefies, maggots eat away the woody parts of the plant, leaving the desired bark.

8. Pliny, *Natural History,* 12.30, 32.

9. Pliny, *Natural History,* 12.14.7, xii.32, xii.35.

10. See Jack Turner's argument in *Spice: The History of a Temptation* (New York: Alfred A. Knopf, 2004). Turner identifies various uses for scents throughout the ancient and medieval worlds.

11. See Marcus Milwright's discussion in "The Balsam of Maṭariyya: An Exploration of a Medieval Panacea," *Bulletin of the School of Oriental and African Studies* 66.2 (2003): 199–200.

12. Milwright, 200.

Chapter One

Scent in Civic and Domestic Space

Throughout the Roman and Arab worlds, most people encountered perfumes and aromatics on a fairly regular basis either at civic events or within their own homes. Sweet scents certainly bestowed personal pleasure but they also served a more common public purpose as they communicated the body's social status, gender, and morality. Cultural ideals celebrated "clean" men and women because only the morally bankrupt, lower classes and "others" threatened society with their filth and stench. Romans encouraged men and women to avoid offensive personal odor by bathing, freshening their breath, and trimming their nails.[1] Yet, to display too much care for the body, especially for women, indicated dubious character and sexual promiscuity. Romans should groom themselves—but not too much.

Roman personal hygiene habits indeed indicate basic concerns for the body and its smell. Because they linked excessive body hair with odor, Roman men and women often practiced some form of shaving, depilation, or plucking both the underarm and genitalia.[2] Pliny described underarm deodorant made of iris, alum, or rose petals.[3] Both males and females perfumed their bodies with little mention of gender specific scents. Instead, scent marked the cultivated, urban dweller (a bit like today's metrosexuals?) against the uncivilized rustic.[4] *Too much scent* cast suspicion upon the wearer's character: usually women's chastity and men's virility.

Romans of both genders and all social classes encountered scent at the public baths, one of classical culture's most important staples. Bathing culture allowed Romans both urban and rural, rich and poor, male and female to participate in complex physical and social interactions; by the fourth century, Rome alone boasted 856 baths.[5] The baths usually included exercise areas, hot baths, and cold baths; many even had a special room called the *unctorium* where servants rubbed men and women with scented oils handily stored in bottles located on niches fixed in the walls.[6] The baths' standard aromatic offerings failed to meet wealthier Romans' expectations, however; they regularly brought their own, more luxurious oils and perfumes with them. One mosaic, shown in figure 1.1, depicts a wealthy woman approaching the baths with various servants; one attendant carries a chest, which would have held her toiletry items including strigel (used to scrape oil and dirt from the body), oils, and perfumes.[7]

Some of the baths' waters (especially in hot rooms) and lamps infused the space with sweet smells; flowers and herbs could be added to the water and fragrant oils to lamps to counteract the fuel's acrid odor, a technique also used in homes.

Jews frequented the public baths as well; indeed, many homogenous Jewish communities constructed their own bathing centers across Palestine.[8] Pagans, Jews, and Christians all agreed that bathing brought health benefits, with many baths located at natural springs known for their healing properties.[9] Rabbis did have to address several issues particular to the Jews: first,

Figure 1.1. Roman lady going to baths accompanied by servants, mosaic, vestibule of Roman palaestra, Casale, late 3rd–early 4th century, Piazza Armerina, Sicily, Italy. ***Source: Gianni Dagli Orti/The Art Archive at Art Resource, NY.***

Romans adorned their baths with gods and goddesses, problematic for Judaism's prohibition against idols; second, mixed-gender nude bathing often occurred, impinging upon women's modesty expectations; and, finally, rabbis struggled with whether Jews could bathe on the Sabbath.

One tradition from the Talmud addresses the first issue:

> Rabbi Gamiliel distinguishes form from intent: even though the baths held idolatrous forms (statues), as long as Jews did not worship them (intent), they were allowed to attend the baths.[10]

Archaeological evidence suggests that some Jewish communities adorned their own bathhouses with three-dimensional statues for aesthetic purpose, though others included only mosaic embellishment with fewer figural representations.[11] Ornamentation, even when connected to active pagan worship, did not stop Jews from attending the baths.

This holds true for the final two concerns, modesty and Sabbath regulations. Jewish bathhouses offered alternatives to mixed bathing that threatened traditional notions of modesty. While Roman baths accommodated sexual segregation in the earlier period—i.e., the second half of the second century BCE—mixed bathing became more popular from the first century CE onward.[12] Traditional Jewish expectations of modesty prohibited nudity in mixed groups. Several *halakhot* (Jewish legal statements) required separate baths for men and women even as they lost favor among gentiles. Jewish (and some Roman) baths included special rooms for menstruating women in deference to purity taboos.[13] A form of bathing towel or suit also protected bathers' modesty. With the covering, women and men could avoid complete nudity at the baths altogether.[14] Regarding the final issue of the Sabbath, Jewish communities would extinguish their furnaces for heating water on their holy day or plan their visits to coincide with the Sabbath's end.[15] With these adaptations, Jews effectively incorporated the Roman bathing culture into their own routines.

Attending the baths, for both Jews and gentiles alike, allowed for a fundamentally social experience: people frequented the baths for social interaction as well as personal hygiene and healing.[16] These social interactions included a wide array of services including minor surgery from physicians or dentists; receiving spa services such as depilatories, massages, and dry/wet steam rooms; soliciting prostitutes; and simply having servants cleanse the body.[17] While Jewish rabbis debated mostly the issues listed above, Roman moralists warned against bath attendance because, as a luxurious distraction, it threatened one's virtue and honor. Perfume and scent applied at the baths particularly led to scandalous rumors.

During the era of the Roman Republic, Cicero (d. 43 CE) had criticized men appearing at the Senate redolent with fragrance.[18] Stoic philosopher

Seneca (d. 65 CE) later heralded the "old ways" of heroes such as Scipio, general and statesman of the Republic, against the softer traditions of his own time. He explained that in the more virtuous past, men smelled of "the camp, the farm, and heroism"; men of his own time smelled foul in comparison. "Now that spick-and-span bathing establishments have been devised, men are really fouler than of yore."[19] Pliny complained about the decadence of women bathing nude with men.[20] For such moralists, the clean and sweet-smelling people of the baths, made "soft" by the luxury of bathing, signaled a culture in decline. In contrast, the "natural" smell of virtuous labor—agriculture and politics—recalled a pristine past, a sacrosanct era of Republican morality.

Early Christian writers also expressed concerns about the Roman baths. Clement of Alexandria (d. c. 215 CE) allowed that bathing contributed to good health and hygiene; he did not forbid his followers from attending the baths.[21] He did warn against the lust oftentimes aroused from such pleasurable activities, especially in mixed-baths where men and women gazed upon each other's naked bodies. He repeated the admonition of Matthew 5:28, that looking at the body lustfully equals adultery.[22] Even Augustine (d. 430) discouraged Christians from attending the famed baths at Baiae (south of Naples) because of their "worldly vanities."[23] John Chrysostom (d. 407) lamented that baths made peasants more effeminate.[24] Jerome (d. 420) voiced the most concern about the baths, enjoining women (especially) to avoid them altogether, along with other Roman frivolities such as pastries, wine, and silks. With prolific rebukes throughout his letters to virgins and ascetics, Jerome rated the life of chastity and self-denial as supreme within Christianity.[25] Christian skeptics could not easily dissuade the baths' practical popularity, however; popes continued to build baths situated within church basilicas and monasteries throughout the early medieval period.[26]

Roman, Jewish, and Christian moralists critiqued not only the baths but also other cultural trends deemed both pleasurable and dangerous. While Jews might have accommodated Greco-Roman culture by integrating the baths into their daily lives, rabbis warned against other public activities, including the theaters and arena.[27] These spaces—argued Jewish scholar Josephus, various rabbis, and Christian theologians such as Tertullian—promoted violence, gambling, and other spectacles that threatened Jewish and early Christian morality.[28] Roman gentiles nonetheless frequented these public events with gusto. Roman authorities used scents and aromatics in crowded spaces, which might first suggest an attempt to counteract the overpowering stench of blood, dirt, and sweat in the coliseum. To "refresh" those in attendance, slaves circulated bowls of perfumed oils or scented waters. At theater events attendants sprinkled fragrant water over both the audience and performers.[29] Romans thus employed fragrance within groups, certainly to mask offensive smells, but also, perhaps, to promote a sense of solidarity or unity, functioning beyond individual pleasure or purity. A shared sensory experience unites a group more powerfully than any other while also revealing important power hierarchies.

During the Roman Empire, emperors as well as elite Romans lavishly displayed their power and authority in public venues. Incense burned on ceremonial fires always accompanied the emperor and his insignia to distinguish his position of (sometimes divine) authority.[30] At public spectacles subsidized by emperors and the wealthy donors, crowds (particularly at the Coliseum) entered a type of "social contract" with the ruling elite—the donors provided entertaining shows and the people responded with applause and support, if pleased; or shouts, malicious humor, and even riots, if not. The arena established a vertical social hierarchy by segregating the elite from the nonelite by space and visual cues. Wealthy citizens and their sons sat apart from freedmen, slaves, and women; the citizens wore togas and their sons donned apotropaic phallic amulets (*bullae*) to protect them from the evil eye.[31] Even though this separation existed, the arena also appealed to the body's complete sensorium, which all attendees shared.

One can imagine the cheering crowd's roars and various food venders hawking their goods; the smells of bodies pressed together only slightly masked by perfumed sprays of water mingled with the stench of recreational death; and the visual action of the fights ringed by the emperor, vestal virgins, and senators—Roman society's most powerful members. The sounds, smells, tastes, and visual stimuli bound the group together while always drawing attention to who provided the entertainment and who received it.[32] While elite Romans denigrated the poor for becoming slaves to the body and its delights, there existed a fine line between disciplined pleasure and being ensnared by irrational lusts.

Many early Christian leaders criticized the Roman spectacles' sensory indulgence as they warned their audiences to avoid such physical gratifications. By the fifth century, public games and theaters had long been abandoned by the Merovingians in Gaul.[33] In North Africa, Augustine described his friend's moral destruction as he became enslaved to the displays of the arena—he first heard, then saw, and then became intoxicated with savagery that wounded his soul.[34] Lactanius (d. c. 320 CE) insisted that the Roman public delights threatened the believer's purity and offered nothing virtuous in return. He cautioned that the arena's pleasure, or *voluptas*, polluted the viewers' conscience just as if they had participated in the killing itself.[35] Clement of Alexandria also criticized the luxurious Roman habits and cautioned that sensory fulfillment could threaten salvation.[36]

Because the body's sensorium provided various opportunities for the soul's damnation, many early Christians juxtaposed their own self-control and abstention against the pagan body's revelry and passions. Some hagiographers even inverted Roman somatic ideals and developed a new sensory discourse extolling (semi)nakedness, stench, and fasting. S. Symeon Stylites (d. c. 459 CE) for example, refused to bathe and denied his flesh to the point of physical decay.[37] Holy men and women also deserted the cities, and all of their public pleasures, dedicating themselves instead to solitary existence. Such sacred rhetoric completely upended Roman authority's vertical hierarchy and socially "pure" body.

Paradoxically, Christian teachers also offered Christian martyrs as substitutions for pagan spectacles. The apostle Paul himself had declared that God made followers of Christ "as though sentenced to death . . . a spectacle to the world, to angels and to mortals" (1 Corinthians 4:9). Recounting in impressive detail martyrs' tortures and sufferings became part of Christian liturgy by the fourth century. It might at first seem odd to parallel Christian "sacred" sufferings with the "secular" gladiatorial combats; however, the two share many properties.[38] First, both appealed to voyeuristic passions by focusing on the body, its humiliation (and conquests), and its mutilation. Spectators and listeners alike experienced various deaths in oftentimes erotic detail. Second, even though professional gladiators existed at the bottom of Rome's social ladder, Romans esteemed them for their bravery, virility, and noble deaths. When becoming a fighter, gladiators swore oaths to allow their bodies to be beaten, penetrated, and bound.[39] This voluntary act earned the respect of Romans from all social strata. In similar ways, most Christians also voluntarily offered their bodies for sufferings, not only in imitation of Christ but also mimicking Imperial forms of noble death and suicide.[40] The humble Christian modeled submission to Roman administrators' tragic rulings, merging images of humiliation with victory and glory.

Augustine, for example, who had virulently denounced the games, directed Christians to examine the martyrs' sufferings instead of pagan entertainment; he even heralded God for providing "superior spectacles." In exchanging the arena for the martyrs' suffering, martyrologies' authors strove to reenact the drama for the listener as completely as possible. Just as the Roman public spectacle appealed to the entire sensorium, with arousing sights, scents, sounds, tastes, and touches, so too did the martyrs' deaths. According to her hagiography, Christ Himself proffered Perpetua, a second-century Christian martyr, a tasty treat in a vision preceding her death; then, back in prison, she joined a more substantive "Love Feast" typically provided for the condemned.[41] She also experienced a vision of her brother, Dinocrates, dirty, smelly, and trapped in dark surroundings, ravaged by thirst. After her intercessory prayers, Dinocrates emerged, clean and healthy, with thirst quenched.[42] Perpetua's Christian colleague, Saturus dreamed of a paradisiacal abode awaiting them after death which resembled a rose garden.[43] Martyr texts certainly revealed the body's final sufferings and death, but they also appealed to the listener's sensory imagination throughout the entire narrative. Polycarp (d. 167), for example, suffered first from fire and then by a dagger to his side. When set aflame, the fire vaulted around his body, revealing a form like bread baking in an oven, redolent with sweet smells of frankincense and other spices.[44] Christian spectacles, much like their Roman counterparts, presented the observer (and listener) with visually stunning exhibitions appealing to all of his or her senses.

Roman authority and its publication in the body's sensorium appeared not only in civic venues such as theater and games. Formal banquets afforded another semipublic space for encountering a variety of aromatics, resulting in a more intimate yet equally effective hierarchical display. Roman banquets often included the host (patron) with his dependent clients seated (or, reclining) around him in a carefully orchestrated seating order.[45] Wealthier hosts frequently furnished their guests a bath, complete with fragrant oil rubs before the meal. If unavailable hosts supplied water or oil to anoint the body, hair, or feet.[46] Patrons greeted their guests with incense and flower garlands to wear around their necks or head; reclining couches received a sprinkling of aromatic water; servants spread saffron over floors; and, at the end of the meal, attendants again burned incense or herbs to freshen the air. In one extreme example, Emperor Nero installed shifting panels and water pipes so that garlands and fragrant water sprinkled his guests after their meal.[47] Perhaps mimicking this, satirist (and Nero's courtier) Petronius (d. 66 CE) describes fictional character Trimalchio's dining room, complete with a shifting ceiling that allowed a frame to descend among the diners while supporting perfume containers as party favors.[48] Roman hosts, whether freedmen, citizens, or emperors, flaunted their social status not only by the space provided—furniture, serving ware, adornments, and even shifting ceiling panels—but also the quality of perfumes anointing that space.

At home most Romans likened fresh air with good health, thus they scattered living areas with flowers (especially lilies, narcissus, and roses) and sprinkled furniture with fragrant water (scented with lemon verbena, for example).[49] Pious families offered garlands and incense at the *lares* altars (or, domestic gods).[50] Entryways into the home also could be anointed with perfumed oil, powerfully marking the transition from the public to private.[51] Bedroom doors could be adorned with garlands and walls anointed with fragrant oils. Clothing chests included dried herbs, and scented water and flowers infused beds.

Many Romans spent time ensuring sweet scents in the bedroom because they promoted good health as well as aided in sexual stimulation and conception. Attendants regularly festooned marriage beds with fresh flowers and perfumes just as the lover in Proverbs described to his beloved: "I have perfumed my bed with myrrh, aloes and cinnamon. Come let us take our fill of love till morning."[52] Romans linked the fragrant myrtle particularly with sexual desire; women (of even the lower classes) offered myrtle garlands to the erotic goddess, Venus.[53]

Scent also aroused sexual excitement when directly applied to the body. At the baths and at home, daily hygiene included perfuming the hair, beard, feet, neck, and breasts (or any combination thereof), oftentimes with several different scents. The Hebrew Bible describes one extreme beauty regimen as Esther prepares to meet the Persian King: for six months attendants anointed her with oil of myrrh, and six months with "sweet odours" (Esther 2:12). The Song of Solomon celebrates scent throughout the text, describing the Lover's hands dripping with myrrh and the Bride smelling of nard, saffron, calamus, cinnamon, frankincense, and aloes.[54] Because of this sexual association throughout the Roman world, authors encouraged young unmarried persons to avoid scent and use plain olive oil to anoint their bodies.[55] Rabbis too worried that Jewish males wearing scent, especially when traveling in an area known for homoerotic activity, would be tempted into sexual sin. The Talmud even stressed that males should not venture into the public markets with fragrance on their clothes, hair, or body in case they become engaged in pathic intercourse.[56]

Sweet scents suffused marriage rituals, perhaps because Romans linked fragrance with sexuality, but also because they simply provided pleasure. Romans marked various holidays, festive celebrations, and vigils with alluring aromas that distinguished time and space. At weddings, the bride's kin (if financially able) showered friends and family with flowers and perfumes at their house—the primary rituals of Roman marriage included leading the veiled bride to her new home, accompanied with torches and, sometimes, bawdy songs. Before departing the natal home, guests feasted among censers burning incense and fragrant torches. Brides received unguents to anoint their face and hearts as well as woven myrtle garlands for their hair.[57]

Jewish marriage rites also incorporate fragrance in various ways. One early practice explained in the Geonic law code *Halakhot Gedolot* (c. 589–1038 CE) describes the presentation of the blood-stained sheet after the marriage night.[58] In a public ceremony, a special blessing called *birkat betulim* is recited over wine and spices with the proof of bridal virginity. The consecration first acknowledges He "who creates the fruit of the vine" and "who creates fragrant trees" as well as "placed the walnut in the Garden of Eden, the lily of the valley, so that no stranger shall have dominion over the sealed spring; thus the loving doe preserved her purity and did not break the law."[59] In this ritual performance, wine and spices commemorate God's creation as well as the bride's purity, recognizing the transformation between virgin and bride, the "walnut" now promising fertility and growth.

Other Talmudic associations between bride and scent are far less poetic. According to tradition, wives wear perfume so they will not repulse their husbands. Beginning with the marriage ceremony, a woman's appearance becomes central to her role in pleasing her groom. The bride should be perfumed, decorated, and laden with fine jewelry. The Talmud teaches that throughout the marriage, "A wife is only for beauty, a wife is only for children; and Rabbi Hiyya further taught: A wife is only for [the wearing of] a woman's finery."[60]

In all these encounters with aromatics, as with other sensual pleasures, ancient authors warned against excess. Because smell ranked as one of the "lower" senses, associated more with base impulses, it proved particularly dangerous. Enjoying or donning too much scent signaled a weak mind and dubious morals. Political critics often labeled their opponents as corrupt, irrational, vain, and effeminate because they wore an excessive amount of perfumes. Many authors blamed the decay of Roman culture on excess in general. Epicurean teacher Lucretius (d. 55 BCE) warned Roman males against lust and desire, which inevitably led to a family fortune squandered on perfumes, fine foods, jewelry, and a life ill spent in luxury and debauchery.[61] Pliny includes perfumes and unguents as pure extravagance because, unlike land and movable wealth, they could not be inherited. This made them a temporary distraction, quick to fade away.[62] Christian polemicists continued in much the same vein, condemning the Roman's attention to the body as they linked sensual pleasures with religious depravity. Tertullian (d. 240), for example, associates physical adornment, including perfumes, make-up, and fine dress, with deception and evil.[63]

Scent signaled female morality in particular. The Babylonian Talmud criticized women who

> put myrrh and balsam in their shoes and walk through the marketplaces of Jerusalem, and on coming near to the young men of Israel . . . kicked their feet and spurted it on them, thus instilling them with passionate desire like the serpent's poison.[64]

Women wearing too much scent posed a serious threat; they compromised their own virtue while at the same time, they menaced innocent men who passed within smelling distance.[65] Pliny clearly identifies women, decadent by nature, as the origin of the spice trade and perfume industry. Because of their degenerate fashions, men squandered their time, talent, and treasure on such meaningless acquisitions.[66]

Despite the various critiques, perfume and aromatics permeated life's daily routine. Perhaps more importantly, they also marked the final rituals performed by Romans, Jews, and Christians at death. Romans and Christians (for a time) practiced both cremation and inhumation, although burial became almost exclusive by the fourth century; Jews only buried their dead.[67] Perfumes, flowers, and fragrant ointments pervaded funeral preparation, with cinnamon especially prominent.[68] Indeed Pliny notes that Romans employed so much of southern Arabia's aromatic exports in death rituals that they called the region "Happy Arabia" (*Arabia Felix*) not because it was blessed by gods above but by gods below.[69] Christian poet Prudentius (d. c. 413) noted that Christian mourners wrapped their dead in violets before burial and then sprinkled perfume on the funerary stones.[70]

Most basically, funerary attendants or family members washed the body in fragrant water; this constituted the most fundamental ritual completed even for the poor. Women often prepared the corpse, at once associating them with the most polluting task imaginable as well as the most important role in the death drama. Assuming some level of wealth or status, families then anointed the body with spices and oils. During this process, sweet-smelling smoke fumigated the entire space, symbolically purifying it of both disease and evil. For Jews the body should be prepared at home.[71] For non-Jews, cremation pyres included flowers and herbs. Bodies readied for both burial and cremation could be embalmed either internally (most commonly observed in Egypt and considered in Rome as overly ostentatious) or externally (which included the body's anointing).[72] According to John 19:38–42, Jews applied a mixture of aloe and myrrh over the body and then also wrapped it in a linen shroud.

Roman bodies sometimes would be displayed for several days before inhumation or cremation so, practically, aromatics masked the stench of decay.[73] Perfume jars and aromatics in censers, lamps, and incense burners usually appear with bodies placed in tombs or caves. These items might have been used by whomever prepared the body for burial, suggesting disposable perfuming implements; or, they might be funerary gifts.[74] Perfume bottles represented in memorial mosaics suggest the latter.[75] Funerary scents also advertised the family's wealth, status, and reverence for the dead, just as large funeral wreaths do today, while freshening the air. Mourners considered bad smells not only as unpleasant but also as harbingers of disease (according to contemporary theories of contagion) and beacons for demons and other malevolent spirits. Perfumes and incense thus lent powerful protection against physical sickness and spiritual contamination. Symbolically, the aromatics sustained the deceased's soul and thus needed continuous maintenance along with food and libations.[76] Many Romans' wills subsidized food, incense, and floral libations (especially roses) at their burial sites.[77] Christians interpreted the scent offerings not as items the soul might use in the afterlife but as evocative of Heaven, where the soul would (hopefully) arrive.

Before the final cremation or burial, mourning women again became a focal point of funerary spectacles. Female family members and hired mourners often wailed and ululated, scratched their faces, and dirtied their hair and clothes.[78] Funerals evolved into lavish and ostentatious displays of wealth, including the most expensive perfumes available; the Romans even tried to legislate against such costly display.[79]

In Jewish death rituals, the ritual use of aromatics especially in the post-Temple period, signaled a shift in theologies about resurrection and afterlife.

Even though eschatological teachings are not consistent in ancient Judaism, Diaspora Jews especially displayed an evolving interest in resurrection after death.[80] Those living outside Judea, for example, could disinter their dead, strip the bones of remaining flesh, sprinkle the bones with herbs and incense, and return the remains to their Holy Land for second burial.[81] Even if lost in transit, the bones qualified as a sweet odor to God, assuring their final retrieval from a liminal existence after death to an eschatological reconstitution in the restored Israel.[82]

Rabbis do provide instruction for the body's preparation and burial. Anointment with oils and spices occurred first and then the general washing, which is quite curious; this indicates that Jews included scent for symbolic purpose because certainly if they intended unguents to conceal the smell of death, the corpse should have been thoroughly washed first. After anointing and rinsing, attendants shrouded the body and prepared it for procession. Passages from the Mishnah indicate that incense and spice also could be brought before the procession, especially when a "surplus" of funds was available.[83] Thus scent seems to play no apotropaic role; instead, it becomes part of the sensual display along with the wailing women and clappers. Finally, a larger number of *unguentaria* have been found at burial sites throughout Palestine with relative silence about the practice in rabbinic literature.[84] Perhaps this indicates that, like their Greco-Roman neighbors, Jews simply desired burial with their personal belongings, particularly their perfumed ointments.

Christians, on the whole, continued the rituals concerned for preparing the dead—bathing and anointing the body—although many Church leaders argued against excessively expensive funerals.[85] The Biblical disciples rebuked Mary for the costly ointments she applied to Jesus foreshadowing His imminent death.[86] Likewise, after Peter's martyrdom in the apocryphal *Acts of Peter*, the apostle appears to his follower Marcellus and criticizes him for the extravagant burial he commissioned.[87]

The most egregious funeral displays belonged to the Roman emperors and later, Christian leaders and saints. Emperors' bodies often required embalming when they died outside of Rome or Constantinople. While alive, emperors entered into their capitals among their adorers with a formal ritual called the *adventus* which incorporated elaborate incense and aromatic diffusion.[88] At death, Romans welcomed their emperor in similar fashion, with the carefully preserved body on display for all to see.[89] Before Christianity, some bodies then underwent *apotheosis*, or divinization. During one bizarre ritual, the Imperial form's wax replica, filled with frankincense and spices, replaced the corporal form, signifying that the body itself had risen to heaven among the gods.[90] A more standard option included a funeral pyre, covered with aromatic incense, with a trap door located beneath it. As the pyre burned, the door opened and an eagle—representing the emperor's soul—flew to the heavens in symbolic *apotheosis*.

Christianity assimilated many of these death rituals to their own purpose. Recently dead Christian emperors and Church leaders such as bishops and saints received similar reverential displays with incense marking not the *adventus* into an earthly city but the *transitus* (or, transition) to Heaven. Romans embalmed on site the emperors that died away from the capital to preserve the body during the trip back to Constantinople.[91] S. John Chrysostom, Archbishop of Constantinople, died in exile and received the same tribute.[92] After S. Symeon the Elder's death, monks continuously burned incense; visitors at Emperor Justinian's funeral noted the dizzyingly strong scents present in the room.[93] As the incense burning displayed honor and reverence for beloved saints and Christian emperors, it also dramatized the theological principle that they ascended to Paradise to rest in God's presence.

During the Merovingian period, Church leaders also gradually assumed primary roles in the funerary drama traditionally assigned to kin groups. With Christianity's proliferation, dealing with the dead now included a concern for corporeal preservation for a very real, very physical resurrection. As several burial inscriptions attest, Merovingians consigned the dead to sepulchers and coffins where the physical body would remain until Judgment Day.[94] The death drama also included handling dead body parts as saint veneration spread; holy relics became an indicator of power throughout Gaul. This physical contact with the dead contradicted established Roman and Jewish traditions relating death with impurity. Perhaps ameliorating such concerns for pollution, Church leaders and hagiographers renovated disgusting corporeal remains into fragrant tokens of salvation and piety. The Church identified holy individuals by their uncorrupted and fragrant forms when disinterred. Church authorities recognized these occasions as miraculous, even if bodies had originally been wrapped in various spices. Gregory of Tours, for example, noted that attendants anointed Radegund's body with aromatics at death, even though he did not name which ones—this did not interfere with her sweet-smelling sanctification later.[95]

Many Christian families who had the requisite means for various levels of embalming modeled the burial of their own dead on the exempla set forth by the saints.[96] At one end of the spectrum, the body could be laid on a bed of herbs with flower garlands around the head. At the other end, the dead could be washed in and injected with various preservatives. A skeleton dating to the fifth century found at Saint-Victor in Marseilles was covered in spices and herbs, including incense and thyme.[97] A grave at Saint-Denis held a body identified as Queen Aregund that had been partially embalmed with a solution that covered her corpse and filled her mouth.[98] The presence of spices and sweet smells did not always indicate sanctity. Gregory of Tours, for example, described the body of a young girl he observed in a damaged sarcophagus as sweet-smelling; he also noted that this was a successful attempt at preservation instead of any miracle.[99]

The use of fragrance in Merovingian burial rituals in some ways resembles their use during the Roman Imperial era; here, too, wealthy families provided for the most exotic scents at death as displays of piety but also status. In other ways, however, early medieval burial rituals function very differently, especially regarding the afterlife. Whether families were concerned for corporeal resurrection or desired to imitate the scents of the saints in heaven, many Christians devoted their wealth to funerary scents. In the process, the Church hierarchy became increasingly responsible throughout the Merovingian and early Carolingian period for legitimate burial as they provided sanctified soil and, eventually, commemorative masses that also included lavish displays of incense.[100]

The relative value of the spices and incense employed by the Merovingians was also much greater than those of the Roman Empire. The decline of Roman authority and the proliferation of Christianity in the west had heralded shifts in both use and availability of aromatics in general. Demand particularly for incense waned as Christian leaders at first forbade its use in religious ceremonies as pagan practice not to be copied. Christians only allowed the ritual use of incense within sacred ritual after Constantine's Edict of Milan in 313 CE which legitimized the religion. Until that time Christian liturgists associated fragrance with polytheistic rituals, particularly emperor worship, and condemned its use. Only as Christianity evolved into a state cult did it integrate incense into the liturgy perhaps as a way to compete with its various pagan counterparts; and as the urban areas of the Empire transformed into more rural enclaves, the Church and new aristocracy provided the primary market for aromatics. Ironically, as the Church warned against the lewd Roman baths and terminated theatrical performances, the circus and other public spectacles, the mass and Church ritual provided the primary public occasions for sweet smells.

Unfortunately for the Christians' renewed interest in incense, east-west trade definitely declined by the late fourth century. Without the annual journeys of Roman fleets brought about by the disintegration of empire, Arab middlemen controlled most of the spice trade, and prices soared. Aromatics were certainly present, only more restricted to the new ruling elite—royal families from emerging "barbarian" kingdoms and the Church hierarchy (oftentimes the same family lines). The most prominent presence—and power—of smell resided first and foremost in the ecclesiastical sphere instead of the earlier Roman Imperial hierarchy. Bishops traveling to Church councils such as the one at Nicaea in 325 CE, for example, were guaranteed lodging and provisions including meat and vegetables with spices.[101] By the eighth century, monasteries such as the one at Corbie (in Gaul) received annual taxes that included oil, olives, and various spices such as pepper and cinnamon.[102] The same could not be said for the laity: domestic access to spices and aromatics in general grew increasingly out of reach.

SCENT IN THE ARABIAN MILIEU

Like the Romans before the economic and cultural upheavals of the fourth century, Arabs also employed perfumes and other aromatics throughout their routine domestic lives; however, their domestic milieu differed much from (especially) the Roman Empire's urban areas. In pre-Islamic northern and central Arabia, life was mostly nomadic: independent tribes traveled in seasonal rhythms throughout a largely desert terrain. No public baths, arenas, or gymnasia defined public space. Southern Arabia, more centralized politically, did boast ruling kingdoms and more agriculturally sustainable lands; yet, a fundamentally tribal culture united most of Arabia. The southern region supported the ancient forests of incense and myrrh trees, today greatly diminished both in size and variety. These forests' desiccation resulted from a combination of forces: first, during the incense trade's heyday, many landowners probably over-cropped their trees, forcing numerous harvests for greater profit; second, when trade declined by the late fourth century, planters might have turned their incense and myrrh crops, more like shrubs than trees, over to grazing land for camels and goats; and third, unpredictable droughts and climate changes might have contributed to the withering away of arable soil.[103]

Historians have generally assumed that because European markets prized incense, particularly, in east-west trade, Arabia's economic and cultural foundations must have also revolved around it. It seems, however, that incense crops—and their diminishment—directly affected only a few landowning families while most "ordinary" Arabians occupied their time with agriculture or animal husbandry. Yet, Arabs across economic and social divisions enjoyed a variety of aromatics within both domestic and public life.

Unique forms of social rankings existed in pre-Islamic Arabian society. Individuals identified with their immediate kin, their clan (*qawm*), and finally their tribe (*qabila*), with various levels of wealth and prestige appointed to each. One could be born into a poor family within a relatively powerful clan, just as the Prophet Muhammad; this meant a modest material existence, but one comparatively secure because of your clan's protection. As with the Roman Empire, high-quality spices and aromatics distinguished the wealthier and more powerful families within clans and tribes; yet, cheaper items such as incense played a part in almost everyone's life.

Arabians employed scent domestically in much the same ways as Romans. They regularly fumigated their homes (a common mixture being frankincense and thyme) as well as their bodies.[104] Ninth-century Islamic scholar `Ali b. Sahl Rabban al-Tabari even referred to a common practice of crushing musk and ambergris, then dusting it inside pockets and sleeves for a pleasant personal odor.[105] We have a good idea of the types of aromatics used by south Arabians in particular because of a number of small incense altars, usually between 5 and 10 cm high, with letters (from early S. Arabian alphabet) designating the incense burned in each.

Surprisingly, altars with the letters LBNY—related to the Arabic *luban*, meaning frankincense—seldom appear, suggesting domestic rituals included it less frequently. Archeologists have found other types of incense more often, such as LDAN (ladanum, an oleoresin extracted from the rock rose); QLM (calmus, a scented reed, perhaps lemongrass); KAMKAM and DARW (cancanum and tarum, probably the gum and wood of the mastic tree); and QUST (costus, extracted from a North Indian herb).[106] South Arabians also reportedly used storax, a fragrant gum imported from Syria and Asia Minor, to fumigate their clothes and homes as well as expel snakes from frankincense groves.[107] Arabs often sprinkled aromatics around their homes, especially on cushions and beddings, to enhance the space. Wealthy Arabs even used bits of musk, by far the most expensive and valued scent.[108]

Most Arabs' daily toilet included fragrance. Muhammad stressed the importance of sweet-smelling breath and encouraged the use of aloeswood in toothbrushes—even better if soaked in rosewater.[109] He also recommended freshening the mouth (even licking a perfumed cooking dish) after eating onion and garlic; even angels avoided people with bad breath.[110] Early Islamic sources describe the various perfumes available for the clothing and body; we find much more information about scent and perfume relating to men than we do to women. Extant perfume manuals, for example, provide more recipes for men's perfume compounds than women's. Women certainly wore perfume, but its appreciation should be primarily from the husband. Men, on the other hand, sported a larger variety of complex recipes in multiple public venues.

It seems ironic then that a female's fragrance appears as one of the most important tropes in Arabic poetry; scent effectively signals a woman's sensuality. Pre-Islamic poets link women with fragrant musk more than just about any other sense, including vision. With Islamic expectations of modesty and sexuality, however, came a critique of women who smelled sweet to others than her husband. A woman's fragrance should be reserved for her spouse; when a woman's scent could be detected by other men, it called into question her chastity. A perfume mixing vessel (*madaf*) was a regular part of the *wife's* possessions; she used it to mix scents that would please her husband.

Men's scent, on the other hand, could be appreciated more publicly and marked social and economic status, especially among the Umayyad and `Abbasid courts and aristocracy. Their public venue required more varieties of complex recipes as discussed in perfume manuals. Men regularly anointed their beards and mustaches as well as their bodies and garments. This anointing could occur in private or at the end of a meal. Some scents were decidedly "masculine," including musk; aristocratic men might dissolve musk in rosewater and sprinkle it on their bodies.[111] Other scents were associated only with females, although women wore musk and other "masculine" scents as part of their perfume compounds (or, *ghaliyah*). Distinctly feminine scents included cloves, camphor, and violet and jasmine oils.[112]

According to the early histories, then, women might wear more "masculine" scents; for women, the *strength* of scent and venue led to scandal. Too much scent—or too public—indicated sexual immorality. Men, however, should never wear feminine scent, or too much. This might lead to gender ambiguity, which proved especially dangerous for men. In the Umayyad period, for example, as the center of Islamic rule solidified in Damascus, a privileged class emerged in Mecca and Medina—made wealthy by the pilgrimage trade—that supported an artistic culture. Two of the most important expressions—poetry and song—not only transformed pre-Islamic styles but also concerned religious leaders because they suggested luxury and moral decay. Critics labeled one important group of Medinan entertainers *mukhannathun*, or effeminates.

We know about the *mukhannathun* from several prophetic hadith; therein, they serve primarily as intermediaries between women and men, facilitating in marriage contracts. The Prophet never supported this lifestyle as, according to several hadith collections, he "cursed effeminate men (*al-mukhannathin*) and mannish women."[113] Identified by their languid style, feminine dress (such as saffron-colored robes), colored hands and feet (with henna), and scent, early Umayyads classified the effeminates as neither eunuchs nor homosexuals. Many sang and performed in Medina, at least until a particularly harsh reprisal by (perhaps) Caliph Sulayman (r. 715–17).[114] According to some reports, the Caliph had several (or all) of the Medinese transvestite musicians castrated because of the threat they posed to women's morality—not only did they have regular access to female quarters but their music also roused feminine desire, threatening society's very moral fabric.[115]

Sources remain relatively silent about the *mukhannathun* until they reappear in Baghdad at the `Abbasid court, where aristocrats generally demoralized them as profligate entertainers and shamed them as passive homosexuals. Both female performers (*qiyan*, generally slave girls) and male singers (generally freemen) often cross-dressed for their royal audience. The women (*ghulamiyat*) cut their hair and wore boy's clothing; many even painted side curls and mustaches with perfume compounds in their transvestitism.[116] The males shaved their beards and donned feminine attire, scent, and other accoutrements.[117]

Unlike the Umayyad entertainers of Arabia, the `Abbasids viewed *mukhannathun* as passive recipients in homosexual intercourse (the active participant was known as the *luti*). Indeed homosexual identity appears in public discourse quite frequently throughout `Abbasid literature. Contemporaries even noted the semantic shift, including Basran literary author al-Jahiz (d. c. 868). Al-Jahiz suggests that as `Abbasid armies of eastern Iran ventured out on long campaigns, they turned to male subordinates for sexual gratification. This predilection became more mainstream as it spread westward into Iraq.[118] Whatever the cause, the `Abbasid literary revival celebrated homoerotic poetry and prose; and, notably, it did not garner the religious contempt that might be expected from conservative Muslim leaders. Instead, critics grouped homosexual intercourse among the general sins of promiscuity and sexual licentiousness—for both the effeminates (the penetrated, distinguished by their dress, scent, and carriage) and their lovers (the penetrators).[119]

Literary and religious leaders alike had no qualms celebrating sexuality within religiously sanctioned unions; and Arabs understood that sweet smells on the body and within sleeping quarters enhanced sexual pleasure. Much earlier, Herodotus, a fifth-century BCE Greek historian, had noted that wealthy Babylonians sat over burning incense to scent their bodies before sex.[120] Umayyad poet Kuthayyir suggests that women continued the tradition by allowing the sweet smells under their gowns.[121] When Caliph `Umar (r. 717–720 CE) consummated his marriage with Fatima (a cousin), he mixed a costly compound scent (*al-ghaliya*) in his bed chamber's oil lamps on his wedding night.[122] Sweet smells infused the room and stimulated desire for sexual intercourse. According to Abu Yasir al-Baghdadi, cited in a sixteenth century literary work on scent, "Perfume is one of the greatest pleasure of mankind and one of the strongest incentives to copulation and the gratification of one's desire."[123]

Even though there remains few sources that speak to the routines of pre-Islamic life and gender expectations, we know significantly more about the early Umayyad period, particularly its bathing and hygiene habits. As the Islamic Empire expanded after the Prophet's death, urban and dynastic cultural patterns replaced the desert nomadic tribal systems as the dominant Imperial realities. Muslims conquered the lands of Syria and Palestine, for example, and they inherited the bath culture left by the Romans. Even though baths had been on the decline in the western Empire, such as in Gaul, the Eastern world continued to support them well into the Islamic era. Although the Byzantine world suffered under plague and population decline during the sixth and early seventh centuries, bath culture remained vital, only scaled down with more private and monastic patrons (verses Imperial) paying the bills.[124]

Public baths appealed to the Arabs, thus they continued using them but in a slightly different form: the large complexes were replaced by smaller, more modest baths containing (in most cases) hot waters, steam rooms, massage rooms, and lounging areas where servants provided cool drinks and fans.[125]

The public bath, or *hammam* thus emphasized the pleasures of hot water and massage only (without the cold waters of the traditional Roman baths),

Figure 1.2. Harun al-Rashid and the barber. ***Source: Ascribed in notes to Bihzad and to Mirak. Public domain. Made available courtesy of the British Library.***

as do traditional Turkish *hammams* today. Semitic custom also did not allow for mixed-bathing, strictly segregating the sexes. Men and women either frequented completely different structures, or they alternated days.

Early Islamic *hammams* offered a variety of services including (with the earliest) blood-letting, massages, and hair care. The duties of the "barber" (*muzayyin*) included depilatory use, removing the body hair from underarm and pubic area, in accordance with Islamic tradition.[126] Men and women (and often servants) would have brought their own toiletries, including any perfumes, massage oils, and henna for dying the hair.[127] By the early `Abbasid period, a common way that women cleansed their hair included a type of clay infused with rosewater or orange flower water. Owners of the bath also fumigated the bath with scent at least twice a day.[128] Some Islamic scholars criticized the baths because they saw them as innovative trappings from a foreign culture, and others saw public nudity as an affront to Islamic expectations of modesty. Soon, however, scholars (generally) accepted the baths as an Arab practice, solved the nudity problem by requiring loincloths for all attendees, and discouraged excessive attendance, especially for women (see figure 1.2).[129]

Certain controversies surrounding the baths did evolve among Islamic teachers and particularly legal theorists. Although present, the vigorous moral suspicion presented by some Jewish, Christian, and Roman authors did not abound, perhaps because Muslims never completely separated the baths from religious ritual. Indeed, baths stood adjacent to mosques and incorporated Islamic codes of conduct. Worshippers completed the ablutions required in Islamic law that returns the body to a pure state (or, *wudu'*). Minor impurities—such as urination and defection—call for limited ablutions of the arms, lower legs, and orifices of the face; major impurities, however—such as sex and menstruation—require full immersion (or *ghusl*), practically met at the *hammam*. Purity regulations required full immersion particularly for women ending their menstrual cycle; the Prophet also directed women to purify their genitalia with musk.[130] In addition to these circumstances, Islamic law recommended ritual baths for Friday prayers, special celebrations such as Ramadan's end, and pilgrimage.[131]

One important point of controversy concerned the water's purity at the baths. Legal scholars disagreed as to whether waters defiled by physical detritus or even morally polluted persons negated the immersion ritual. Some hadith suggested that the *hammam* might offer health benefits, but not ritual purity. According to one tradition attributed to Ibn Abbas (d. 692), believers should cleanse again outside the *hammam* to qualify for ritual purity.[132] Scholars also prohibited the sacred rituals at the baths because believers should neither repeat the name of Allah nor read from the Qur'an in such unclean space; they even likened the *hammam* to cemeteries.[133] Others maligned the baths as the house (*bayt*) of Satan and the jinn (especially at night).[134] By the tenth century or so, these juridical concerns had been mostly quieted and bath attendance for ritual cleansing and physical pleasure no longer posed a legal quandary. Like the Rabbinic Jews, Muslims assimilated many aspects of classical bath culture while fitting them uniquely to their own social and sacred purpose. The *hammam* thus afforded social interaction, especially a relatively "public" feminine space, and pleasant sensual experience while at the same time conferring religious purity.

As in Rome, scent played an important role in everyday life, not only at the baths but also festivals and life rituals, especially those surrounding death. What we know about pre-Islamic funerary rituals comes mostly from archaeology. In south Arabia (around the region of Yemen today), unique funereal rites suggest there may have been a belief in an afterlife. Various grave offerings proliferate throughout the sites, including amphorae and personal items (such as jewelry); by the sixth century BCE, rudimentary embalming even appears.[135]

South Arabians also practiced two distinct forms of burial—multiple and individual—with the latter giving way to the former by the fourth century BCE. Multiple burials include the turret tombs, located some distance from any settlement, along the major trade routes. These tombs might have belonged to organizers of caravan trade.[136] Other multiple graves were located in rock and cave tombs. Here, archaeologists have found abundant examples of miniature versions of real-life artifacts including tables, altars, weapons, and incense burners. Some burials, especially individual, contain residue of sacrificed meat (of sheep or goat) and camels (buried within or near the grave). Whether this designates belief in an afterlife or not, mourners placed items both familiar and customary within the burial space, with incense burners prominent among them.

Here again, we know much more about funerary practices in the early Islamic era, especially from hadith collections and legal texts. As Islam spread out of the Arabian Peninsula and encountered various cultural systems, death rituals assimilated local practices. Some general patterns do emerge, however, especially by the ninth and tenth centuries. It is clear that aromatics infused both public and private funerary rituals. Same-sex family members and sometimes funereal professionals prepared bodies for burial. They routinely burned candles and incense in the washing rooms neutralizing death's stench. Some standard cleansing methods included washing the body three, five, or seven times with water containing ground leaves from the Lote-tree.[137] Attendants then anointed the body with camphor. These particular traditions relate back to prophetic hadith; according to legal scholar Malik (d. c. 795), the Prophet gave these instructions to women on how to prepare his daughter Zaynab's body after her death.[138] Generations of Muslims later imitated the Prophet's injunction, including camphor whenever possible in preparing bodies for burial.[139] This was not always available, however, as camphor originated from resinous East Asian and Indian trees.

Washers also applied camphor, a quite pungent scent, to cotton and then patted the corpse's mouth, nostrils, and deep wounds. They placed remnants of the scented cotton between the buttocks to contain any effluvia that drained from the body on the way to burial. Some jurists stressed that the main corporeal areas of filth, including the groin, armpits, and fold of knees, be cleansed and scented. Most importantly, musk should be applied to the forehead, palms, and knees, the human body's most dignified parts, the areas of prayer prostration.[140] Muslim men and women generally took the responsibility of providing for their own burial expenses very seriously; they deemed the burial shroud as the most important item which could also be perfumed or imbued with incense.[141]

As Islamic traditions concerning death rituals evolved, they effectively differentiated Muslims from their Jewish, Christian, and Zoroastrian counterparts in important ways. Unlike the Jews (and the Romans), for example, Islamic tradition transformed the handling of dead bodies into a pious deed instead of an act of pollution. Those Muslims who prepared the dead incurred no ritual impurity and, indeed, were not even required to perform ablutions after their service.[142] As within Christianity, Islamic tradition virtually reversed the general sense of ritual disgust associated with death. Islamic burial rituals also emphasized beautifying the dead, perhaps preparing the body for resurrection. To these ends some jurists allowed for the postmortem removal of pubic hairs and even circumcision, both customary in many Islamic cultures.[143]

Unlike Christian tradition, however, Islam emphasized the body's full dignity and integrity. Even though Muslims associated holy men and women with sweet smells even after death, they never allowed for the division and distribution of dead body parts among the pious. Ibn Ishaq (d. 767), for example, reports in his *Sira* that, according to the Prophet's cousin and son-in-law Ali ibn Abi-Thalib, Muhammad's body still smelled sweet after death.[144] However, it would have been unthinkable to promote his corporeal relics' adoration. Notions of sanctity certainly existed in Islam as well as the veneration of holy men and holy women, although the ritual adoration and cultic proliferation of dead body parts never flourished as within Christianity.[145]

The expansion of the Christian cult of the saints in the West signaled the Church's burgeoning power; the locus of religious authority became increasingly identified with the prevailing ecclesial hierarchy. For the history of scent, this means that the Church displayed that authority by appealing to the worshipper's full sensorium in mass, holy day festivities, and rituals relating to saint venerations—including the use of incense and perfume no longer widely available to "average" Greco-Romans and Merovingians. In the Islamic world, however, we do not find a parallel decline in the availability of scent and spices among lower classes and their centralization among a clergy. Instead, we find ever more conspicuous—and extravagant—displays of scent among religious leaders as well as the Umayyad and early `Abbasid courts.

While scent certainly played an essential role in domestic and civic performances, Arabian literary culture also relied upon it for symbolic purpose. From pre-Islamic classical poetry to Umayyad and `Abbasid court propaganda, authors employed the sensual in portraying love, peace, as well as power. Compared with value placed on incense in the Roman world, we might expect to find it referenced often in literature; on the contrary, Arabs never esteemed incense (even the highest quality) so highly above other aromatics. Tenth-century poet al-Sari al-Raffa' compiled an anthology of popular poetic references to lovers and beloveds, scents, and drinking (*al-Muhibb wa al-mahbub wa al-mashmum wa al-mashrub*); he discusses musk, ambergris, camphor, and aloeswood, but does not even mention frankincense or myrrh.

By the fifth and sixth centuries, a type of aromatic canon had appeared in Arab literature, boasting five central fragrances that demonstrated wealth, power, and status. This canon included musk, extracted from the male musk deer in central and eastern Asia; ambergris, harvested from the intestinal tract or the sputa of sperm whales; aloeswood imported from India and Southeast Asia; camphor, a resin from camphor trees found in Asia; and saffron, used as an aromatic or a dye, imported from Fars, Syria, and Spain.[146] A somewhat later source, al-Suyuti's (d. 1505) *al-Maqama al-miskiyya*, a clever literary form that described four embodied perfumes contending for first place among the scents, declares musk as the winner because of its various religious and medical purposes.[147] Most of these scents, introduced into the Arabian Empire through Sasanian Persia, remained unknown in the classical Roman world.[148] The procurement and distribution of these relatively new and exotic scents signaled the Arabian owners' status and lavish lifestyle.

Classical poetry, spanning the pre-Islamic period into the Umayyad era, details this "aromatic canon" most poignantly. Arabian culture has always esteemed poetry and venerated poets; some traditions even suggest that divine beings inspired the best poets, bestowing upon them a sublime skill unmatched by human effort. Patrons employed poets and paid them for poetry fit for various occasions. Several authors compared themselves to the pestle and stone that grinds perfume. The poet "grinds the musk" by creating verses which waft the patron's sweet smell and honor.[149] Within the thousands of extant verses, cultural historians glean information about religion, gender, tribal law, and nature through set meter, rich in symbolism and metaphor; social historians also uncover clues about tribal identity, social rank, and daily life (primarily of the aristocrats). Scent encodes all of these categories—usually connected with a high-ranking and sexually provocative female.

Poetic master Imru al-Qays (d. c. 544 CE), for example, uses spices and perfumes to summon ancient odes' (or *qasida*) most important themes: nature and women. In his *Mu`allaqa* he stares across a deserted settlement and compares the gazelles' dung with peppercorns, an imported and expensive commodity.[150] When recalling his past lovers, he describes their strong, musk scent that "wafted from them like breath of the east wind bearing the fragrance of cloves."[151] Poet `Antara also compares his lover's breath to a merchant's musk bag.[152] These images present striking metaphors because of musk's exotic nature and elevated rank among the perfumes. By engaging the world through these most glamorous scents, the poet boasts of his status and sexual prowess.

In al-Muraqqish al-Asghar's ode, *Qasida mimiyya*, censers burning sweet aloeswood greet the beloved Bint `Ajlan each evening in her tent.[153] Fumigating domestic space in this way appears popular throughout Arabia. Even though both men and women perfumed their bodies and rooms, the poets more often associate sweet smells with women, usually their past lovers. This might simply be because of perfume's symbolic qualities—sweet and ephemeral—like the lovers' memories themselves.

Poets generally describe women and perfume in their poem's beginning, called the *nasib*, which offers a sentimental remembrance of the poet's beloved. Islamicist and literary scholar Michael Sells has pointed out that the poet relies upon the full sensorium to generate a subjective image of the beloved (often described as a lush garden), including sight (water, clouds), taste (wine, fruit), touch (wet, warm), and smell (musk, spices).[154] Seventh-century poet Duh al-Rummah, in "To the Encampments of Mayyah" (*a manzilatay mayyin*), describes his beloved:

> She reveals/Petals of chamomile/cooled by the night/to which the dew has risen at evening/from Rama oasis,/Wafting in from all sides with the earth scent of the garden, redolent as a musk pod falling open.[155]

Throughout the poem, Duh al-Rummah relies mostly upon smell, touch, and vision to recall for the hearer his beloved's nature. Aromatic imagery, along with other sensuous allusions, thus function as a literary strategy, building into an orgasmic crescendo of remembered pleasures, ending finally in a language of ablution and purification (usually, with images of water).[156] The final images of water (here the oasis) after a series of sensual allusions, might also insinuate Semitic purity practices after sex.

Literary scholar Suzanne Stetkevych has suggested another reading of both classical odes and a second classical literary genre—female-authored elegies (or, *marathi*)—by equating perfume with blood.[157] Stetkevych argues that both poetic forms embed a ritual structure of sacrifice and redemption. The ode usually begins with the *nasib*, where the poet grieves over a deserted settlement while musing over his lost love(s); an arduous journey then follows, culminating in his boastful tribute to himself and his tribe, usually with an animal sacrifice and celebration. The elegy parallels this structure of departure and sacrifice, leading to rejoicing and redemption, by opening with a lamentation of a beloved kinsman's death (sacrifice) followed by the calling for blood vengeance (redemption). Linguistically, the root of battlefield (عرك) relates to menstruation (عركت), linking the two forms of blood loss. Within the elegy, the poetess decries a particular kind of blood loss—her kinsmen's unavenged death—as an impurity that requires ritual action. Metaphorically, retribution on the battlefield returns the tribe to a pure state, just as water and perfume return a menstruating woman to purity.[158] Indeed, the root of perfume (*tib*) can also denote purification, gladness, or sweetness. As Stetkevych points out, poets often shamed men who failed to avenge their kinsmen's death by comparing them with women who had not purified after menses.[159]

In one short elegy attributed to al-Khirniq bint Badr ibn Hiffan, for example, the poetess says:

> Let my kinsmen not be distant! / Men who are the enemy's poison, / The slaughter camels' bane, / The attackers on every battleground, / The perfumers of / Their loincloths' knots.[160]

In classic elegiac form, al-Khirniq calls on men of her tribe, men deadly to their enemies, to avenge her fallen kinsmen which include her husband, son, and two brothers.[161] Stetkevych also reads "perfumers [*al-tayyibuna*] of their loincloths' knots" to signify men who perfume or purify themselves with the blood of vengeance. In this case, men enact ritual purity by steeping their loincloths—tied at the actual point of their manhood—in their enemy's blood just as women bathe the genitalia in water and perfume after menstruation.

Classical authors also associated perfume with pacts and covenants of peace as well as warfare. Ibn Ishaq (d. 767) relates perhaps the most famous example of this in his *Life of Muhammad (Sirat rasul allah)*, as transmitted by Ibn Hisham (d. 833), when he describes the truce achieved between two Qurayshi families (the Prophet's own tribe) in pre-Islamic Mecca. According to Ibn Ishaq, various families quarreled over land and responsibilities within Mecca, the center of Qurayshi control. Two strong families emerged, those related to and allied with the Banu `Abd al-Manaf (the precursors to the Prophet's clan) and the Banu `Abd al-Dar. The Banu `Abd al-Manaf ritually enacted the loyalty with their allies by sending for a bowl full of perfume and reciting an oath as they and their supporters stirred their hands within the bowl.[162] Family leaders then spread their perfumed hands over the *ka`ba* (or, great house)[163] to further seal the pact; they became known as the "scented ones" (*al-mutayyibun*).[164]

Pre-Islamic poet Zuhayr's famous ode (*qasida*) also evokes this meaning as he describes tortures of war and promises of peace. In his *Mu`allaqa*, he commends tribal mediators' efforts at ending a long-standing feud when he writes:

> You alone mended the rift between Abs and Dhubyan / after long slaughter, and much grinding of the perfume of Manshim, / and you declared, "If we achieve peace broad and sure / by ample giving and fair speaking, we shall live secure."[165]

The "perfume of Manshim" might refer to the pacts concluded, establishing peace instead of war. The sweet smells of conciliation replace the blood loss of constant warfare.

With this reading in mind, we recognize perfume's multivalent imagery within Arabian poetry and culture in general; poetry provides more than just an historical laundry list of scents available in pre-Islamic Arabia. At a symbolic level, poets discuss perfume and sweet smells most prolifically at the beginning of the ode/elegy, thus associating it with departure and death within the ritual structure. According to ritual theory, this is a liminal phase, a transition from one state (absence/death) into another (redemption/celebration), impurity into purity. Sweet smells signal such transitional moments in most rites of passage—such as death and marriage—so it follows that poets would here employ their most fragrant imagery. The poetic presence of perfume announces the transition from purity to impurity, loneliness to love, death to redemption, war to peace.

Classical poetry continued as an important literary artifact throughout early Islamic history, but an urban court culture replaced the Arabian Peninsula's nomadic tribal system that produced a plethora of literary genres and histories. The powerful Umayyad family established a ruling dynasty (or caliphate) centered in Damascus, poised at the center of a new and expanding world empire; as such, they acquired incredible wealth and prestige. Ostentatious displays of riches, which included expensive scents and perfumes, occurred on scales well beyond the means of any pre-Islamic tribe.

Gift exchange among and between rulers certainly had been practiced in the pre-Islamic Middle East. One eleventh-century text, surviving through a fifteenth-century edition, recounted some of the most opulent displays of royal and dynastic wealth. The *Book of Gifts and Rarities* (*Kitab al-hadaya wa al-tuhaf*), for example, describes the lavish gifts from India and China to Sassanid (Persian) emperors in the late sixth century, just before the advent of Islam.[166] These treasures included ornate silks, gems, and scent—most often musk and camphor. Both rulers introduced themselves to the Persian emperors by describing their vast lands and possessions; their homes and palaces, they claimed, fragranced vast distances.[167]

After Islam proliferated throughout the Middle East, Indian and Chinese rulers transferred their conversations, negotiations, and grand displays of friendship to the Caliphs and their administrators. One of Caliph Hisham's (r. 724–743 CE) governors received an extravagant gift from an Indian king which would probably strike a modern viewer as garish instead of particularly royal. The king sent a she-camel figurine, studded with precious stones, mounted on wheels that rolled. When the governor later forwarded it to the caliph, Hisham pierced its udders, from which flowed pearls "like milk," and its neck, from which fell blood-red rubies.[168]

Not only do these Indian gifts appear brash to modern sensibilities, however. In the ninth century, Caliph al-Ma'mun married his daughter Umm al-Fadl to Abu Ja`far Muhammad b. `Ali al-Rida (considered the ninth Imam by many Shi`ite Muslims). At the wedding feast's start, servants pulled in, by silk ropes, a large silver boat filled with various expensive perfumes. They then anointed all esteemed guests' beards, and then perfumed the "commoners."[169] And when a respected general and ambassador invited Caliph al-Muqtadir (r. 908–932) to his farm, he reportedly sealed the water's flow into a local stream, packed it with ice, and then "made water and drinks flow into them." He prepared a picnic by lining the stream's banks with adorned baskets, hanging various roasted meats from trees, and surrounding the seating areas with musk, camphor, ambergris, and saffron.[170]

The reports of such gifts, of course, might be exaggerated, but the emphasis and care directed to gift-giving is not. Islamic tradition distinguishes between two different types of gifts—first, *hadiyya*, usually requires a mutual exchange of gifts. This implies the participants' relative equality in status; indeed, the gift exchanges recorded in the *Books of Gifts and Rarities* occurred between rulers, administrators, or royal families. The Prophet enjoyed *hadiyya*, encouraging his followers to engage in mutual gift-giving, comparing it (symbolically) to the sharing of love and friendship.[171]

Gift exchange, especially between rulers, often appears as a display of one-upmanship instead of affection. When the Byzantine emperor sent Caliph al-Ma'mun an elaborate gift, the Caliph instructed his courtiers to: "Send him a gift a hundred times greater than his, so that he realizes the glory of Islam and the grace that Allah bestowed on us through it." The caliph then insured its grandeur by adding a large quantity of musk and 200 sable pelts.[172] With these exotic exchanges, the caliph advertised both his power and wealth—the musk, indeed an expensive gift, and the sable, originating probably from Russia or China.[173]

Gift-giving that distinguished asymmetrical power relationships exists in most political systems—including early medieval Gaul; it is not unique to early Islamic dynasties. Early Christian texts referred to gifts as *munus* or *donum*; the *munus* in particular signaled the gift giving practice among unequal partners. For example, S. Augustine explained that the Holy Spirit's presence in a Christian's heart as a *munus* from God; likewise, pious Christians made their gifts or offerings to God as tribute.[174] Such gift-giving, effectively forging an intricate web of loyalty and obligation, filled early medieval Gaul's political machinations. Scent or perfume rarely appears among the gifts given, however. When it does, it certainly indicates wealth and status; but, in early medieval Gaul, gift exchange more likely included horses, weapons, precious metals, gems, or even precious books.[175] Scent and perfume, more often associated with ecclesiastical power, slipped in both supply and relative symbolic value among the Franks.

A second type of gift-giving in Islamic tradition, called *hiba*, emphasizes such hierarchical status; it is voluntary and without expectation of reciprocation. This type of gift-giving occurs between a subordinate and a ruler, in order to display one's allegiance or good wishes. An Iraqi administrator, for example, sent Caliph Hisham (r. 724–743) a collection of clothes and perfumes for his household as a display of allegiance and, perhaps, gratitude for political favor.[176] Some actions of *hiba* did not end well, however. In one case, Caliph al-Mutawakkil's (d. 847–861) concubines presented him elaborate gifts at a public festival. His beloved (and favored) Shajar, a slave girl, included twenty tamed gazelles wearing saddlebags filled with musk, ambergris, and fine scents. The caliph clearly prized this gift among the others, which led to the jealous concubines to plot Shajar's murder.[177]

Acts of public *hiba* can be the reverse, however (i.e., rulers and the wealthy presenting gifts to favored individuals, guests, or the public in general). This resembles early medieval Gaul's gift-exchange practices between Frankish royalty and ecclesial authorities and their clients. When Harun al-Rashid (r. 786–809 CE) married Zubaydah, he arranged a display of wealth, generosity, and charity as then unmatched in the Islamic era. He of course presented her with costly personal gifts including precious stones, scents, servants, and maids; but his generosity extended "to the people" as he distributed dinars and dirhams from silver bowls, containers of ambergris and expensive mixed scents (*ghaliyah*), and robes of fine fabric.[178]

Scent and aromatics thus functioned in several important ways in Roman and Arab public/private lives at the time of nascent Christianity and Islam. Daily routines and popular bath culture reveal standards of personal hygiene which were far from private affairs. Indeed, the baths provided opportunities for social interaction as well as sexual scandal. Public baths in Islam linked also with religious purification, shielding them (for the most part) from critiques similar to those advanced by Christian moralists. Most rites of passage incorporated scent, including both marriage and death rituals. Christianity and Islam associated dead, holy bodies with sweet scent (discussed more in Part

II) and looked forward to a fragrant Paradise. Attendants anointed Christian corpses with flowers or unguents anticipating their resurrection; Muslims scented their dead as well, paying special attention to the five points of prostration (knees, palms, forehead). Public gatherings, festivals, and commemorations—and the recognition of heroes and holiness—relied upon scent to bind the audience in a shared sensory experience. Yet, men and women remained vulnerable to public criticism if they wore "too much" scent or the wrong kind; indeed, outward aroma reflected inner virtue.

Ecclesial and dynastic courts demonstrated their emerging power and status with fragrant spectacles and gift-exchange. Gift-giving publicly featured the hierarchical patron/client relationship that effectively formed social bonds. In Europe, the Church evolved as the primary storehouse for aromatics and scent, distributed as they intended. In the Islamic world, dynastic courts reveled in their expanding power with ostentatious displays of sensory indulgence. Sumptuous meals and banquets often stood at the center of these occasions, and food rituals included their own fragrant etiquette. We now turn to Roman and Arab gustatory habits that further demonstrate their unique sensory worlds.

NOTES

1. See Kelly Olson's discussion in *Dress and the Roman Woman: Self-Presentation and Society* (New York: Routledge, 2008), 70. Pliny also notes that some Romans trimmed their nails on market days beginning with the forefinger for good luck. *Natural History*, 48.48.

2. Body hair generally provides a breeding ground for bacteria which causes odor. For the ancient world, see Olson, 65–66. Herein, she discusses various sources that mention hair removal including Seneca, *Letters* 56.2; Ovid *Art* 1.506; Pliny, *Natural History*; among others. Many Romans deemed women without body hair to be most attractive, and many women removed their genital hair in preparation for sexual activity. See Pliny, *Natural History* 26.164 and Lucian, *Amores*, 26.

3. Pliny, *Natural History* 21.142, 35.185, 21.121; discussed in Olson, 70.

4. See Olson, 78. As she notes there are exceptions; some authors felt that no men should be scented. Martial also identifies balsam as a "masculine" scent (14.59) while Juvenal says the opposite (Juv 2. 41–2).

5. Caseau, 134.

6. Petronius describes Trimalchio getting ready to leave the baths, as being soaked in scented oil and dried with fine linens. *Satyricon* §28. For a fine archeological collection of perfume bottles, see Mikhal Dayagi-Mendeles, *Perfumes and Cosmetics in the Ancient World* (Jerusalem: Israel Museum, 1993).

7. Piazza Armerina, Sicily; discussed in Garrett G. Fagan, *Bathing in Public in the Roman World* (Ann Arbor: University of Michigan Press, 2002), 195.

8. See Yaron Z. Eliav, "The Roman Bath as a Jewish Institution: Another Look at the Encounter Between Judaism and Graeco-Roman Culture," *Journal for the Study of Judaism* 31.4 (2000): 416–54.

9. See Estēe Dvorjetski's *Leisure, Pleasure and Healing: Spa Culture and Medicine in Ancient Eastern Mediterranean* (Boston: Brill, 2007). Dvorjetski includes rabbinic discussions of hot baths as healing in chapter 5.

10. *M Avodah Zarah* 3.5.

11. Eliav, 434–36.

12. See the concise review of scholarship concerning segregated and mixed bathing in Roy Bowen Ward, "Women in Roman Baths," *Harvard Theological Review* 85.2 (1992): 125–47.

13. *M Ketub*, 7.8; *tosefta* Nid. 6.15; *M Nid.* 7.4; discussed in Eliav, 443–44.

14. For references regarding such a garment, see Yehoshua Brand, *Keley zekhukhit besifrut ha-talmud* (Jerusalem: Mosad Ha-Rav Koon, 1978), 176–77, as discussed in Eliav, 445–46.

15. Eliav, 449–50. Jewish communities also maintained separate bathing pools for ritual purity; their construction followed a complex set of Talmudic requirements. These pools, or *miqwaot* (s. *miqweh*), were located at the synagogue (especially during the Second Temple Period) or at private homes. These ritual pools transferred purity to men, women, and ritual implements after immersion. See Ronny Reich, "The Synagogue and the *Miqweh* in Eretz-Israel in the Second-Temple, Mishnaic, and Talmudic Periods," and Zeeve Safrai, "The Communal Functions of the Synagogue in the Land of Israel in the Rabbinic Period," in *Ancient Synagogues: Historical Analysis and Archaeological Discovery, v. 1*, eds. Dan Urman and Paul V. M. Flesher (New York: Brill, 1995), 289–97 and 181–204 (respectively).

16. Fagan discusses the healing benefits of public baths (not just mineral springs); he notes that Pliny discusses various drugs that can be used with baths for healing; e.g., "lupin flour kneaded with vinegar and smeared on in the bath" helps with pimples and ulcers. Celsus, in contrast, focuses more on bathing as a preventative measure along with a good diet. See "Bathing for Health with Celsus and Pliny the Elder," *Classical Quarterly* 56.1 (2006): 195.

17. See Fagan's *Bathing in Public in the Roman World;* also, Estēe Dvorjetski, *Leisure, Pleasure and Healing: Spa Culture and Medicine in Ancient Eastern Mediterranean* (Boston: Brill, 2007), 121.

18. Cicero, *In Catilinam Oratio* II.3, 5, as discussed in Béatrice Caseau, *Εὐωδία: The Use and Meaning of Fragrances in the Ancient World and Their Christianization (100–900 AD)*, unpublished dissertation, Princeton University, 1994, 128–29.

19. Seneca, *Epistulae morales ad lucilium* 86.9–12, trans. Richard M. Gummere; *LOEB Classical Library*, 3 vols. (Cambridge, MA: Harvard University Press, 1917–1925), 2, 315–19. Cleostrata, a *matron* from Plautus' *Casina*, also criticizes her *senex* husband for his perfume; she explains that conflict with Roman civic values. See Catherine Connors, "Scents and Sensibility in Plautus' *Casina,*" *The Classical Quarterly, New Series* 47.1 (1997): 305.

20. Pliny, *Natural History* 36.121, 33.153.

21. For a discussion of Roman views on bathing and health, see Garrett G. Fagan, "Bathing for Health with Celsus and Pliny the Elder," *Classical Quarterly* 56.1 (2006): 190–207.

22. Clement of Alexandria, *Paedagogos* 3.5. See also Eliav's discussion, 424, and Ward 142–43.

23. Augustine, *Contra academicos* 2.26 (Knöll, *CSEL* 63.27); discussed in Eliav, 424.

24. John Chrysostom, *In Act. Ap.* 18.4.

25. See particularly Jerome's, *Epistulae* 45 and 22. Jerome also notes that baths are unnecessary within monastic life; see *Epistula* 14*: Ad Heliodorum monachum* 10 (Hilberg, *CSEL* 54, 60), where he notes: "He who has bathed once in Christ does not need to bath again." As discussed in Eliav, 424.

26. For examples, see *Liber Pontificalis*, Hilary (461–468) and Symmachus (498–514).

27. See Martin Jacobs, "Theatres and Performances as Reflected in the Talmud Yerushalmi," ed. Peter Schäfer and Catherine Hezser, *The Talmud Yerushalmi and Graeco-Roman Culture*, v. 1, 327–47.

28. Tofseta *'Aboda Zar.* 2.7, as discussed in Eliav, 425–26; Tertullian, *De spectaculis*, 21.

29. See Caseau 142–43.

30. Herodian, I. 8, 4. See Caseau, 163–65.

31. See David Fredrick's discussion of seating divisions in "Mapping Penetrability in Late Republican and Early Imperial Rome," in *The Roman Gaze: Vision, Power, and the Body*, ed. David Fredrick (Baltimore: John Hopkins University Press, 2002), 243–53.

32. See Jerry Toner's argument in *Popular Culture in Ancient Rome* (Malden, MA: Polity Press, 2009), especially chapter 4, "Common Scents, Common Senses."

33. See in particular Yitzhak Hen's discussion of "Merovingian Secular Culture," in *Culture and Religion in Merovingian Gaul, 481–751* (New York: Brill, 1995).

34. Augustine, *Confessions* 6.8.

35. Lactanius, *Divinae Institutiones* 6.20; see also Toner, 160, and Edwards, 58.
36. Clement of Alexandria, *Stromata* 7.7.36.
37. See Susan Ashbrook Harvey's "On Holy Stench: When the Odor of Sanctity Sickens," *Studia Patristica: Papers Presented to the International Conference on Patristic Studies* 35 (1998): 90–101.
38. See Catherine Edwards' discussion in chapter 7, "Laughing at Death"; and Carlin A. Barton, "Savage Miracles: The Redemption of Lost Honor in Roman Society and the Sacrament of the Gladiator and the Martyr," *Representations* 45 (Winter 1994): 41–71.
39. See Horace, *Satirae* 2.7.58–59, for one version of the oath. Discussed in Barton, 52.
40. Edwards makes a compelling argument for models of noble death and suicide that dominated the late Republican and Imperial era as the military ideal was no longer available to Roman male citizens.
41. *Martyrdom of Perpetua and Felicitas*, 17.
42. *Martyrdom of Perpetua and Felicitas*, 7–8.
43. *Martyrdom of Perpetua and Felicitas*, 11–13.
44. *Marytrdom of Polycarp*, 15.2.
45. See J. H. D'Arms, "Control, Companionship and *Clientela*: Some Social Functions of the Roman Communal Meal," *Echos du Monde Clasique* 289 (1984): 327–48; and "Performing Culture: Roman Spectacle and the Banquets of the Powerful," in eds. B. Bergmann and C. Kondoleon, *The Art of Ancient Spectacle* (New Haven: Yale University Press, 1999), 301–19. See also, Fredrick, 253–59.
46. In his *Odes*, Horace promises his friend Maecenas rose garlands and balsam oil to anoint his hair should he visit, iii.29.
47. Suetonius, *Lives of the Later Caesars*, Book VI, Nero, 136–37.
48. Petronius, *Satyricon*, §60.
49. Pliny, *HN*, 25, 59, 107; Caseau, 119.
50. *Odes*, iii.23. Horace suggests that offerings of rosemary and myrtle substitute for animal sacrifice. He also describes offering "incense with prayers" at a domestic shrine to Venus, i.30.
51. Also, Lucretius, *De rerum natura*, 4.1160–1192, describes an amorous man who anoints his beloved's doorpost with oil of marjoram.
52. Proverbs 7:17–18.
53. See Ovid, *Fast*, 4, 139, 143. For a discussion of myrtle's association with sex and desire, see P. G. Maxwell-Stuart, "Myrtle and the Eleusinian Mysteries," *Wiener Studien* 85 (1972): 145–61. Maxwell-Stuart also mentions myrtle's prominence in Jewish rituals, 159.
54. Song of Solomon 5:5, 4:13.
55. Caseau, 127–28; also see L. Holmes, "Myrrh and Unguents in the *Coma Berenices*," *Classical Philology* 87.1 (1992): 47–50.
56. See *b. Ber.* 43b; see Michael Satlow's discussion in "'They Abused Him like a Woman': Homoeroticism, Gender Blurring, and the Rabbis in Late Antiquity," *Journal of the History of Sexuality* 5.1 (July 1994): 1–25.
57. A marriage scene is described in *Acts of Thomas*, 5; also, Claudian, *Epithalamium*, v. 154–55, 208–10. Discussed in Caseau, 158–59. Women's grooming articles and personal adornments were included among the *mundus muliebris*; these items could be legally bequeathed by husbands for their wives. The articles could include mirrors, unguents, and various containers, sometimes made of precious silver, thus potentially conferring significant wealth. See Susan Treggiari's *Roman Marriage: Iusti Coniuges from the Time of Cicero to the Time of Ulpian* (Oxford: Clarendon Press, 1991), 213, 389.
58. See Ruth Langer's fine article, "The *Birkat Betulim:* A Study of the Jewish Celebration of Bridal Virginity," *Proceedings of the American Academy for Jewish Research* 61 (1995): 53–94.
59. As translated in Langer, 53.
60. BT *Ketubot* 59b; discussed in Nissan Rubin's *Time and Life Cycle in Talmud and Midrash: Socio-Anthropological Perspectives* (Boston: Academic Studies Press, 2008), 105–9.
61. Lucretius, *De rerum natura*, 4.1122–1140.
62. Pliny, *Natural History*, xiii.4, xxi.1.1.

63. See particularly Tertullian's *On the Apparel of Women*, 2. Turner, *Spice*, also discusses, 72–73.

64. Babylonian Talmud, Shabbath, 62b; also discussed in Caseau, 127.

65. Also, Lucretius, *De rerum natura*, 4.1160–1192, describes women's deceptive and seductive natures, appearing more beautiful to desirous men than they really are. One aspect of that deception is the excessive use of perfumes.

66. See Pliny's discussions in *Natural History* throughout 11–13.

67. Caseau, 173.

68. Also discussed in Turner, *Spice*, 148–49.

69. Pliny, *Natural History*, xii.41.18.

70. Prudentius, *A Hymn on the Burial of the Dead*, lines 169–72, pp. 105/112.

71. *Semahot*, 1.3; Caseau, 177.

72. Toynbee, 42; Caseau, 175.

73. See Toynbee discussion in chapter 3, "Funerary Rites and the Cult of the Dead."

74. See Caseau's discussion, 185–93.

75. For example, the Tomb of the Valerii under S. Peter's Cathedral; see Toynbee's discussion, 53.

76. In *Satyricon* §71, Trimalchio asks his friends to build a tomb for him with a puppy, garlands, and bottles of perfumes painted at the base so that he can continue to enjoy them as he did in life.

77. Toynbee, 62–63.

78. See Valerie Hope's discussion in *Roman Death* (New York: Continuum, 2010), 125–26. Although, as Hope points out, as the Imperial era progressed, elite women were expected to restrict public displays of emotion; professional wailers and mourners took their place, 128.

79. See Cicero, *De legi*, ii, 24, 62, discussing the early laws of the Twelve Tables; Ovid, *Fasti* vi, 663–64. Discussed in Toynbee, 54.

80. See Lionel Rothkrug's description of Rabbinic Judaism's view of incense offering and the dead in "The Odour of Sanctity and the Hebrew Origins of Christian Relic Veneration," in *Historical Reflections* 8.2 (1981): 95–142.

81. See Rothkrug's description, 102–5.

82. See Ezekiel 37. Some scholars regard Ezekiel's discussion of the animation of "dry bones" as an early allusion to an eschatological gathering of Israelites (a form of resurrection?).

83. See *t. Šeqal.* 1.12; discussed in Green, 57.

84. Green, 61–62.

85. Augustine, for examples, even criticizes the lavish feasts and drinking bouts held at burial sites at annual commemorations of and festivals for the dead. See *Against Faustus*, 20.21 and *Letters*, 22.

86. John 12:1–7.

87. *Acts of Peter*, 36–40. Caseau, 176–77.

88. Horace describes both public and domestic celebrations. *Odes* iii.14 directs a servant to bring garlands, perfumes, and fine wine to celebrate one of Caesar's triumphant returns.

89. Or, in some cases, likenesses of the emperor accompanied by their ashes. Emperor Severus, for example, died in Britain, was cremated with ashes mixed with spices, and then sent to Rome in an urn. See Herodian, III, 15, 7–8, Whittaker, 368–69; Caseau, 168–69.

90. See E. Bickerman, "Consecratio," in *Le Culte des Souverains dans l'Empire Romain*, Entretiens sur l'Antiquité classique, Vandoeuvres-Genève, 1973; Caseau, 168–69.

91. This includes Constantius II, Jovian, and Valentian I. See Caseau, 182–83.

92. *The Syriac Chronicle of Zachariah*, II, 5; Caseau, 183.

93. See *Life of Simeon Stylites*, trans. Lent, 196; and Corripus, *In laudem iustini augusti minoris*, libri IV, trans. Cameron, 22–25. Discussed in Caseau, 183–85.

94. See Bonnie Effros, *Caring for the Body and Soul: Burial and the Afterlife in the Merovingian World* (University Park: Pennsylvania State University Press, 2002), 71.

95. Gregory of Tours, *Gloria confessorum*, 104, 366.

96. See M. E. A. Pigeon, "De l'embaumement des morts à l'époque mérovingienne," *Bulletin archéologique du Comité des travaux historiques et scientifiques* (1894): 139–42.

97. Discussed in Effros, *Caring for Body and Soul*, 74.

98. Effros, *Caring for Body and Soul*, 74.

99. Gregory of Tours, *Gloria confessorum*, 34, 768–69; discussed in Effors, *Caring for Body and Soul*, 73.

100. See Cyril Vogel, "Deaux consequences de l'eschatologie grégorienne: La multiplication des messes privées et les moines-prêtres," in *Grégoire le Grand: Chantilly, Centre culturel les Fontaines*, eds. Jacques Fontaine, Robert Gillet, Stan Pellistrandi, Colloques internationaux du CNRS, September 15–19, 1982 (Paris: Editions du CNRS, 1986), 267–76.

101. See Turner, *Spice*, 84.

102. See Bonnie Effros, *Creating Community with Food and Drink in Merovingian Gaul* (New York: Palgrave Macmillan, 2002), 30; Jean Lestocquoy, "Épices, medicine et abbayes," in *Études mérovingiennes: Actes des journées de Poitiers, 1er–3 mai 1952* (Paris: A. et J. Picard, 1953), 179–86.

103. For an important discussion of climate changes in Arabia, see Groom's chapter 11, "Climate and History in Arabia."

104. For a fine review of modern uses in southern Arabia/Yemen, including fumigation and socialization, see Dinah Jung, *An Ethnography of Fragrance: The Perfumery Arts of `Adan/ Laḥj* (Boston: Brill, 2011).

105. Ṭabarī, *Firdaws* 611. Cited in Anya H. King, *The Musk Trade and the Near East in the Early Medieval Period.* Unpublished dissertation, Indiana University, 64.

106. See Nigel Groom, "Trade, Incense, and Perfume," in *Caravan Kingdoms: Yemen and the Ancient Incense Trade*, ed. Ann C. Gunter (Washington, DC: Smithsonian Institution, 2005), 105–13.

107. Pliny, *Natural History* 12.42.

108. Imru al-Qays, *Mu`allaqa*, 41; trans. in Alan Jones, *Early Arabic Poetry*, v. 2 (Oxford: Ithaca Press Reading for the Board of the Faculty of Oriental Studies, 1996). In one extravagant example, a government official prepares his home for Caliph al-Rashīd by sprinkling musk on everything in the house and the garden (even sweet-smelling flowers) so that the caliph would be completely surrounded by the pleasant (and expensive) scent. *Books of Gifts and Rarities*, §114.

109. Al-Suyūṭī, 69.

110. Al-Suyūṭī, 72.

111. See Aḥmad al-Washshā's manual on aristocrat behavior under the `Abbasids, *Kitāb al-Muwashshā*, ed. R. E. Brünnow (Leiden: Brill, 1886). Discussed in King, *Musk Trade*, 185–86.

112. See al-Washshā, 186–87. Discussed in King, *Musk Trade*, 194–95.

113. See, among others, Ibn Ḥanbal, *Musnad*, nos. 1982, 2006, 2123, 2291, 3458, 7842, 7878, 5649, 5328; al-Bukhārī, 77.103–4. Discussed in Everett K. Rowson, "The Effeminates of Early Medina," *Journal of the American Oriental Society* 111.4 (1991): 673.

114. The musical scene in early Medina and Mecca is described by Abū l-Faraj al-Iṣfahānī in his *Kitāb al-Aghānī* (Cairo, 1323/1905–06). Al-Iṣfahānī provides numerous variants of the traditions, but Sulaymān appears as the culprit among most. See Rowson, 691.

115. See *Aghānī* 4.59f; Rowson, "Effeminates," 690–91.

116. See Everett K. Rowson, "Gender Irregularity as Entertainment: Institutionalized Transvestism at the Caliphal Court in Medieval Baghdad," in *Gender and Difference in the Middle Ages* (Minneapolis: University of Minnesota Press, 2003), 48.

117. See Kristina Richardson's related article, "Singing Slave Girls (*Qiyan*) of the `Abbasid Court in the Ninth and Tenth Centuries," in *Children in Slavery through the Ages*, eds. Gwyn Campbell, Suzanne Miers, and Joseph C. Miller (Athens: Ohio University Press, 2009), 105–18. As Richardson notes on p. 114, the transvestite female performers particularly appealed to Caliph al-Amin's sexual attraction to young males. His mother, earnest for an heir, reportedly hired slave girls to entertain her son dressed as boys.

118. Rowson, "Gender Irregularity," 59.

119. We have a rare opportunity to view Christian criticisms of the "effeminate" Islamic court through the Martyrdom of S. Pelagius (c. 925), written by hagiographer and Cordoban priest, Raguel. In the text, Caliph Abd al-Rahman III, an Umayyad ruler of al-Andalus, apparently invites the beautiful boy-saint to engage in sexual play. Pelagius refuses, telling the Caliph that he is not "effeminate like yourselves" (*effeminatum*). This, of course, reflects the

Christian castigation of Islamic morality but might also speak to the gender diversity within caliphal courts. See Mark D. Jordan's brilliant discussion of Pelagius' martyrdom in *The Invention of Sodomy in Christian Theology* (Chicago: University of Chicago Press, 1997).

120. Herodotus, *Histories*, trans. George Rawlinson (Knoxville, TN: Wordsworth Classics, 1996), Book I, chapter 198.

121. See al-Raffā', v. 3, p. 165, no. 306; also discussed in King, "The Importance of Imported Aromatics in Arabic Culture," 177.

122. *Kitāb al-Hadāyā wa al-Tuḥaf (Book of Gifts and Rarities)*, trans. Ghāda al-Ḥijjāwī al-Qaddūmī (Cambridge, MA: Harvard University Press, 1996), §111.

123. See G. J. H. van Gelder's discussion in al-Suyūṭī's "Four Perfumes of Arabia," 207.

124. See Hugh Kennedy's discussion in "From Polis to Madina: Urban Change in Late Antique and Early Islamic Syria," *Past & Present* 106 (1985): 3–27.

125. See 'ḥammām," *EI2*.

126. One of the most popular forms might have been "sugaring," a process still popular in many baths and salons, especially in North Africa. It involves using a sugar and lemon juice (sometimes honey) as a form of epilation. According to some hadith, the Prophet preferred a lime depilatory (*nura*), made from arsenic mixed with water. The area could then be anointed with henna, vinegar or rose-water. See al-Suyūṭī 116.

127. Several hadith also identify henna as a sexual stimulant; dying hair increases sexual appetite. See al-Suyūṭī 82.

128. Discussed in Ahsan 199.

129. There is an amusing anecdote recorded in a journal during the early crusades written by Usāmah ibn-Munqidh (d. 1188). A "Frank" (a term used to describe all crusaders) enters a public bath operated by a Muslim owner. He tears off the Muslim's loincloth because he expects nudity in the baths. When he observes the bath owner's hairless genitalia, he asks him to remove not only his genital hair but that of his wife as well. Usāmah uses this as an example of the Franks' lack of honor—i.e., asking the male bath owner to observe (and touch!) his wife's nude body. *The Book of Contemplation*, 149–50.

130. Muslim 3.70.

131. See Mohammed Hocine Benkheira's excellent study "La maison de Satan: Le hammâ en débat dans l'islam medieval," *Revue de l'histoire des religions* 200.4 (2003): 391–443. He expands this in two more articles which focus on legal discussions relating to moral order and nudity within the baths. *Revue de l'histoire des religions*, 224.3 (2007): 319–71 and 225.1 (2008): 75–128.

132. Ibn Abī Shayba, *Musannaf*, ed. Shâhin (Beirut: 1995), 102, no. 1156. Discussed in Benkheira, 410–11.

133. Supporters of the hammams responded that as long as the genitalia were covered, it was acceptable to pray and recite from the Qur'an. See Benkheira, 417–19.

134. Benkheira, 423–29. Also see Ahsan, 199.

135. Burkhard Vogt, "Death and Funerary Practices," in *Caravan Kingdoms: Yemen and the Ancient Incense Trade*, ed. Ann C. Gunter (Washington, DC: Smithsonian Institution, 2005), 81–88.

136. A. de Maigret, "New Evidence from the Yemenite 'Turret Graves' for the Problem of the Emergence of the South Arabian States," in *The Indian Ocean in Antiquity*, ed. J. E. Reade (London: Kegan Paul, 1996), 321–37.

137. The Lote tree is prominent in Quranic descriptions of Paradise; its inclusion in funerary rituals suggests the hope for resurrection.

138. Mālik, *Muwaṭṭa'*, no. 592. Discussed in Leor Halevi's *Muhammad's Grave: Death Rites and the Making of Islamic Society* (New York: Columbia University Press, 2007), 53.

139. Bukharī, 23.16; 23.21. An exception to such normative burial expectations was made for Muslims who died on *hajj*, in a state of *iḥrām*.

140. See Halevi's discussion, 65–71.

141. Halevi, 91.

142. See Halevi's discussion, 78–79.

143. Halevi, 65–66. G. H. A. Juynboll also talks about the use of hair dye on corpses; see *Studies on the Origins and Uses of Islamic Ḥadīth* (Hampshire: Variorum, 1996).

144. Ibn Ishaq, *The Life of Muhammad: A Translation of Isḥāq's Sīrat rasūl allāh*, trans. A. Guillaume (Oxford: Oxford University Press, 1997), 687–88.

145. This is a generalization. The veneration of relics does exist in the Islamic world, although without the usual corporal dismemberment usually associated with Christian relics. For example, one of the most popular relics in the Islamic world is a whisker from the Prophet Muhammad's beard, hair he shaved before *hajj*, and his sandal.

146. See Anya King, "The Importance of Imported Aromatics in Arabic Culture: Illustrations from Pre-Islamic and Early Islamic Poetry," *Journal of Near East Studies* 67.3 (2008): 175–89.

147. See G. J. H. van Gelder's discussion and translation in "Four perfumes of Arabia: A Translation of al-Suyūṭī's *al-Maqāma al-miskiyya*, in *Res Orientales* XI (1998): 203–12. The other three scents mentioned are ambergris, saffron, and civet. There existed an earlier debate between musk and civet from the ninth century no longer extant.

148. See Anya King's convincing argument in her *Musk Trade*, chapter 3.

149. See Abū al-Ṭayyib Aḥmad b. al-Ḥusayn al-Mutanabbī, *Dīwān* (Beirut: Dār Ṣādir, 1958. Discussed in King, *Musk Trade*, 148.

150. Imru al-Qays, *Mu`allaqa*, 3.

151. Imru al-Qays, *Mu`allaqa*, 8.

152. See A. J. Arberry's translation, 179, in *The Seven Odes*.

153. Al-Muraqqish al-Aṣghar, *Qaṣīda mīmiyya*, 6, 7; trans. in *Early Arabic Poetry, v. 2*, 246.

154. See Michael Sells, "Guises of the Ghūl," in *Reorientations? Arabic and Persian Poetry*, ed. Suzanne Stetkevych (Bloomington: Indiana University Press, 1994), particularly 140–59. I discuss the garden imagery more fully below in section 3.

155. See Sells' translation in *Desert Tracings: Six Classic Arabian Odes* (Connecticut: Wesleyan University Press, 1989), 70–76.

156. Sells, "Guises of the Ghūl," 140–41.

157. See Suzanne Pinckney Stetkevych, *The Mute Immortals Speak: Pre-Islamic Poetry and the Poetics of Ritual* (Ithaca, NY: Cornell University Press, 1993).

158. See especially, Stetkevych, *The Mute Immortals Speak*, 227.

159. Stetkevych, *The Mute Immortals Speak*, 175.

160. Stetkevych, *The Mute Immortals Speak*, 168.

161. Stetkevych, *The Mute Immortals Speak*, 168.

162. Many historians, both medieval and modern, associate Mecca with the incense trade, asserting that the Qurayshi tribe was central in trade (in general) and perfume (in particular). Patricia Crone questions this in *Meccan Trade and the Rise of Islam* (Princeton, NJ: Princeton University Press, 1987). Crone does confirm, however, that Meccan trade included perfumes, however limited, because Syria and Byzantium had their own perfume industries.

163. This is the same *ka`ba* that plays such an important role in Islamic ritual today—the structure, draped in black cloth, located at the center of the Mecca Masjid al-Haram. The *ka`ba* served as a shrine and a type of "civic center" in pre-Islamic Arabia.

164. Ibn Ishaq, *The Life of Muhammad*, trans. A. Guillaume, §84–85, p. 56–57.

165. See A. J. Arberry's translation in *The Seven Odes: The First Chapter in Arabic Literature* (New York: Macmillan Co., 1957), 115.

166. *Kitāb al-hadāyā wa al-tuḥaf (Book of Gifts and Rarities)*, trans. Ghāda al-Ḥijjāwī al-Qaddūmī (Cambridge, MA: Harvard University Press, 1996).

167. *Kitāb al-hadāyā wa al-tuḥaf*, §8, §29.

168. *Kitāb al-hadāyā wa al-tuḥaf*, §15.

169. *Kitāb al-hadāyā wa al-tuḥaf*, §118.

170. *Kitāb al-hadāyā wa al-tuḥaf*, §126. These adornments resembled the pleasures of Paradise, suggesting the Caliph's piety as well as power. We will discuss these paradisiacal images more in the final chapter.

171. See al-Qaddūmī's introduction, *Books of Gifts and Rarities*, 4.

172. *Book of Gifts and Rarities*, §31.

173. See al-Qaddūmī's notes, *Books of Gifts and Rarities*, 266.

174. Augustine, *Tractate on the First Epistle of John*, 6.9.

175. See Florin Curta's "Merovingian and Carolingian Gift Giving," *Speculum* 81 (2006): 671–99.

176. Curta, "Merovingian and Carolingian Gift Giving," §16.
177. Curta, "Merovingian and Carolingian Gift Giving," §33.
178. Curta, "Merovingian and Carolingian Gift Giving," §111.

Chapter Two

Fragrant Food

Classical and early medieval cuisine usually provided a full olfactory feast: aromatic spices might add flavor, but sometimes scent was added only for scent's sake. Many Arab cookbooks described fragrant waters that could be sprinkled upon completed dishes simply for the diner's pleasure. Both Romans and Arabs also believed that smell stimulated taste, thus adding spices and perfumes aroused one's appetites. The Roman elite class, who had ready access to a plethora of spices and scents, delighted in flavorful combinations. They served both vegetable and meat dishes with fragrant sauces, some containing eight or nine distinct spices and herbs. In the Middle East, recipes described how to scent pots and other cooking utensils as well as perfume foods before presentation. One medieval cookbook includes a list of "basic" spices every kitchen needed—all nine of them (separate from essential herbs).[1]

Extant culinary manuals also resemble nutrition guides, medical manuals, and philosophical (or theological) treaties more than what we today identify as cookbooks. The author of a Roman cookery book, for example, explained that aromatic salts can aid in digestion, serve as a laxative, and prevent all diseases (including colds and the plague).[2] While such salts might cure everything that ails the body, unfortunately the author did not express *how much* salt might be needed; ancient cookbooks thus included a simple ingredient list with no exact measurements.

In early Christian manuals, authors defended their elaborate recipes and ingredients by emphasizing their health benefits—certain food combinations can heal the body, the temple of God. With that function stressed, concerns for gluttony virtually disappear; although, Christian authors also warned that without sufficient care, good food could lead to sexual licentiousness, a grave sin indeed. In a tenth-century Arab cookbook, the author devotes sixteen precise chapters to various foods' distinctive humoral qualities allowing the cook to provide a "balanced" diet for his (and her) guests.[3] With proper knowledge, a good diet could restore harmony and health to the body.

Romans' basic diet included several essential staples: cereals (such as barley and wheat); legumes (such as beans, chickpeas, and lentils); vegetables and fruits; and, less present, meat and fish.[4] Bread probably provided for at least 50 percent of the general caloric intake.[5] These foods cut across socioeconomic lines; how the food was prepared, where it was prepared, and with whom it was eaten distinguished one class from another, and sometimes one ethnicity from another.

This proves particularly true for the Jews living in Roman Palestine. On the whole, Jewish communities shared the same basic dietary staples as their gentile counterparts. The Jewish diet's biblical taboos encompassed most importantly pork and camel; restrictions against pigs distinguished Jews more so because Romans consumed more pork than any other meat.[6] Beginning in the Hasmonean Period (167–142 BCE), Jewish dietary proscriptions evolved stressing, for example, commensality instead of just consumption. The book of *Jubilees* requires Abraham's children to separate themselves from, and "do not eat with," the gentiles.[7] Although some Jews certainly dined with non-Jews, the requirement for social segregation appears more often. Early Rabbinic texts even correlate Jews who eat at gentile banquets with idolaters and assign kosher (allowed) food preparation to Jewish cooks only.[8] Another prohibition required the complete separation of dairy and meat (even at the table).[9] Thus even though Palestinian Jews participated in the general Roman commercial system (e.g., Jews purchased food items at markets), food preparation rituals and commensality effectively defined Jewish identity against their gentile counterparts.

Apicius' *Art of Cooking* provides the best source for understanding Roman culinary practices. The named author, Apicius (if such an individual actually existed), lived in the first century; however, most scholars agree that the extant text dates to the late fourth century and probably contains some of Apicius' original recipes. Cooks and gourmands added to the core collection as time passed.[10] This text continued to play an important role in early medieval kitchens; emended copies survive through a sixth-century cleric named Vinidarius (this version exists in an eighth-century manuscript) and ninth-century monasteries at Tours and Fulda.

Apicius' culinary manual contains some recipes that might fit humble household budgets, probably urban, with access to the diverse ingredients; most, however, describe extravagant dishes that appealed to aristocrats (or, at least, their cooks). The ten chapters present directions for the preparation of vegetables, meats, grains, and various sauces. The recipes required a plethora of herbs and spices that would have been too expensive for most Romans. Among the most commonly mentioned herbs we find mint, oregano, rue, and lovage; popular imported spices include white and black pepper, ginger, cinnamon (or cassia), asafoetida,[11] cardamom, and cloves. The manual listed more customary (and affordable) items such as fish sauce (called *garum* and/or *liquamen*),[12] wine (either straight or boiled down to various consistencies), and olive oil.

Sally Grainger provides modern adaptations of Apicius' recipes in *Cooking Apicius: Roman Recipes for Today* (London: Prospect Books, 2006). These include, most importantly, measurements and contemporary substitutes. In the following salad recipe, for example, she replaces "chicken and goat sweetbreads" with chicken livers.

Salad:

3 1/2 ounces chicken livers, 2 to 3 large livers
1 tablespoon olive oil
half a cucumber
2 tablespoons capers, chopped
1 ounce pine nuts, a generous 1/4 cup
1 to 2 large ciabatta loaves
1/2 cup each: water, white wine vinegar
1 3/4 cups grated Parmesan cheese, about 2 ounces

Dressing:

freshly ground black pepper
1 level teaspoon celery seeds, toasted, see note
3 heaping teaspoons fresh chopped mint
1 ounce pine nuts, a generous 1/4 cup
1 1/2 ounces full-fat cream cheese, softened
2 egg yolks
1 tablespoon each: honey, vinegar, fish sauce
5 tablespoons water

1. Prepare the salad ingredients: Cook chicken livers by frying in a little olive oil, then cool and chop them into small pieces; peel and slice the cucumber thinly; drain and chop the capers finely. Cut the ciabatta into thin slices and lay them out on a large tray. Combine water and vinegar; dribble over the bread slices and allow the fluid to be absorbed. Repeat occasionally while you prepare the dressing.

2. For the dressing, grind the pepper and roasted celery seeds in a mortar or spice grinder. Add the mint and the pine nuts; grind to a fine paste. Add the cream cheese, egg yolks, and the honey; blend again. Then dilute the paste with the vinegar and the fish sauce. Gradually add the water and blend.

3. Take a 2-pint pudding bowl (4-cup mixing bowl); lay a piece of bread, gently squeezed and cut to fit, in the bottom. Sprinkle a little of the liver, capers, pine nuts, and Parmesan over the bread; press down gently. Finish with the cucumber, but do not let the slices overlay each other. Add layer of bread and repeat until all the ingredients have been used up. Always finish with a layer of bread. (Alternatively, you can line the inside of the bowl with bread slices, then proceed.) Pour the dressing over the salad; press down gently. Cover with plastic wrap; chill for 2 hours. Turn out onto a plate; decorate with more cucumber slices.

Overall, the culinary spectrum presented in the Roman cookbook as well as other texts appears extravagant to most modern sensibilities for a number of reasons. First, most foods (especially sauces) contained a wide variety of spices, although spicy flavors might have been ameliorated by quantity. For the majority of these recipes, we just do not know how much cinnamon, pepper, cumin, and saffron went into a dish. We might also imagine that with so many spices and herbs present, it might be difficult for a "general" consumer to distinguish subtle flavors. Perhaps even the smallest inclusion of such varied spices indicated more about accessibility and status instead of gastronomic preference.

Second, cooks often presented food in elaborate ways, hiding one food inside of another (imagine biting into chicken and finding ham and prawn) and even returning feathers to fowl after they had been cooked. Petronius' *Satyricon* mocks these practices as he describes peacock eggs prefilled with fowl prepared in peppered yolk, and hare cooked and then "fitted with wings to make it look like a tiny, furry Pegasus."[13]

Finally, many recipes called for a lavish amount of meat in what might be considered today very odd combinations. Roman cooks layered and stuffed one meat with various other kinds: one of Apicius' patinas (soufflés) includes bits of sow's udder (a particular delicacy), fillets of fish, chicken, turtledove, and "whatever good things there may be."[14] Another recipe called for a pig to be stuffed with chicken, thrushes, sausage, and snails (along with many vegetables and herbs).

Roman mealtime extravagance prompted great controversy, especially during the social and political upheaval between the Republic and Empire.

Moralists romanticized the simpler "early days" of self-reliant farmers, enjoying simple foods, with concomitant moral restraint. For such critics, not only did virtuous Romans smell less like the baths and more like the soil, they also enjoyed a simple diet devoid of gastronomic extravagance. Statesmen such as Cato, Varro, and Columella linked Roman identity with agricultural origins. Pliny even exaggerated the virtues of simple, basic foods such as wheat (not barley, which was associated with the Greeks), chickpeas and beans. These foods symbolized an idealized past and purity, not status and wealth flaunted by expensive, foreign items.[15]

Concerned senatorial authorities indeed tried to regulate extravagant and costly food displays with sumptuary laws; one of the earliest that involves food is the *Lex Fannia*, 161 BCE. The law limited the amount of money spent on any banquet, forbade foreign wine, and stipulated that any dinner should be limited to only a single winged creature not "fattened" just for that meal.[16] The law also restricted the number of guests present at banquets, suggesting a concern for the power relations proliferating between patrons and clients versus the senatorial elite. The law did not succeed; elites and later emperors regularly ignored the sumptuary regulations, evidenced by the various laws that followed, also aimed at curbing private expenditures on meals, also ineffective. The law reminds us that food involves more than a material required to sustain the body but also a symbol of power and status.

This appears most clearly in the disconnect between the textual descriptions of extravagant food consumption and artistic portrayals. When surveying extant mosaics and paintings, between c. 200 BCE and 300 CE, we find that banquet scenes generally focus on the social drama played out in the dining space, including banqueters' seating arrangements, elaborate drinking vessels and serving dishes, and attendant slaves. They do not portray the garishly prepared dishes in exotic poses or adornments we read about. Instead, we find the first depictions of elaborate dishes—including sows' udders—in funerary monuments, featuring nonaristocrats at lavish banquet tables. Within this context, food represented the afterlife's satiety and happiness instead of picturing the deceased's daily routines. Only in the fourth century do house paintings of the elite feature elaborate banquet foods.[17] What this shift in food's visual representation means is open to interpretation. We do know that dining space changed at this time as well: by the fourth century, the *stibadia* (dining couches and tables) had superseded the *triclinium* (couches where diners reclined) in many elite homes.[18] Instead of portraying power through physical location amongst the reclining banquet and opulent serving vessels, perhaps the artist indicated status and wealth with exotic foods, especially as Roman access to spices grew increasingly out of reach in the later Empire.

No matter how difficult Eastern aromatics were to acquire for the elite, however, spices and sweet smells never fully disappeared from Western culinary guides and they continued to distinguish elite class and status. One particularly important manual survives from the early sixth century that reveals how many Roman cookery traditions persisted into the Merovingian era. The letter *On the Observance of Food*, authored by one Anthimus, ambassador of the Ostrogoths, provides a glimpse into the fusion of Roman, Frankish, and (some) Gothic culture and cuisine. Anthimus probably wrote his letter for Merovingian King Theuderic (d. c. 533 CE) while living in northeastern Gaul because he reports many Gallic dietary and social customs.[19] His articulates his noble goal: to provide advice on good nutrition and health, even though that requires him to sometimes criticize the strange (for him) palate of the Gauls.

Anthimus had obviously received some medical training, and he provides more practical advice than philosophical context. Anthimus does not, for example, focus on the body's humoral vicissitudes as outlined by Roman physician Galen (d. c. 217, as discussed in more detail below). He only occasionally refers to the benefit of wet versus dry foods for unbalanced bile or phlegm; however, he repeats the ancient adage in his introduction, "everything in excess is harmful."[20] Anthimus probably did not expect his audience, the Gallic court, to have a background in medical theory; or, perhaps he intended his work to function only as an accessible guidebook.

Some recipes resemble those in Apicius' own cookbook. Wine and honey are often used in various reductions. Cooked fowl appears, although Anthimus does not describe the Roman's elaborate and creative ways of stuffing and presenting it. In fact, Anthimus' tastes seem rather bland in comparison. He favors simple boiling over other cooking means such as roasting, frying, and grilling; for him, boiling insures that meat will be cooked fully and thoroughly.[21] He does include recipes for Roman delicacies such as sow's womb and udder, yet simply boiled or fried.[22] He occasionally mentions spices such as ginger and cloves, but he explains that pepper and wine serve as the best flavorings.[23]

The manual also reveals some new Western gastronomic trends. The Franks' butter and lard replaced the Romans' olive oil; beer replaced wine; and meat became more of a staple. Anthimus assures his readers that "beer is good for anyone" and promises that when "well brewed [it] possesses goodness and surpasses expectation," as if some of his aristocratic readers might need convincing.[24] While he does mention some fish dishes, more of his recipes include meat: beef, mutton, venison, goat, and ox. Even though he stresses the importance of cooking meat, he nonetheless acknowledges the Franks' predilection for eating raw bacon.[25] He offers advice about using milk, much more popular among Franks than their Roman counterparts, in recipes and as a beverage. Anthimus remained a little suspicious of it, warning not to drink it uncooked without some honey, wine, or mead.[26] He also cautions against all but the freshest (and sweetest) cheese.[27]

Anthimus wrote this manual at the Ostrogoth king's behest. That Gaul's neighboring king sent a manual on food as a token of diplomatic relations suggests one of the most important aspects of food: sharing food—even recipes—binds people together. When people eat together, they certainly fulfill a basic need, an obligation that every human being must attend. Yet the rituals associated with food consumption function in various ways: common meals provide an opportunity to rehearse collective memory; display one's power and status; and forge relationships with visible (as well as invisible) others.[28] By delivering a food manual to Theuderic, Anthimus develops a "soft" negotiation among disparate voices. The Ostrogoths and Franks differed on just about every other issue, most importantly Christology: the Ostrogoths adhered to Arian Christianity which the Church had labeled as heresy. By entering a conversation on food, rulers (one viewed by the Church as orthodox and the other a heretic) laid a foundation for more difficult political mediations.

Food rituals—especially among aristocrats and royalty—are charged with symbolic meaning. Like the Romans before them, the Frankish kings (as well as other "barbarian" leaders) often communicated their wealth and authority by hosting grand banquets and feasts; presence at such events dramatized both loyalty and (often unspoken) submission. With Anthimus we do not see feasting between the kings, but we do see an exchange of knowledge about food, drink, and health. In Anthimus' manual, for example, he includes medical advice relating to a healthy diet otherwise unavailable to the Merovingians. While perhaps meant to flatter the king, the manual also features the Ostrogoth's *civilitas* and sophistication compared to the Gaul's less refined habits.[29] After all, they did prefer raw bacon.

Other Christian authors, when writing about food and diet, did not focus on individual recipes or dishes; instead, they warned against becoming too ostentatious. Just as Roman moralists criticized males for esteeming scent, perfumes, and exotic foods too highly, so did Christian censors.[30] These religious authors emphasized the link between lust and food, however. Indeed, Christian theologians identified eating as humanity's original sin; and, that was just the beginning of the problem. According to Clement of Alexandria, women's behavior at public banquets usually led to illicit sex. Jerome advised women in particular to avoid fine food (such as pastries) and wine because it increased their heat and inflamed their genitals.[31] Such calls for asceticism led many Christians to deny themselves rich, spiced food (at minimum) or adopt extremely austere food regimens (e.g., only water, bread, and raw vegetables).

Generally, theologians discouraged severe acts of self-denial—such as excessive fasting—even in monastic communities. Early medieval authors attempted to tame the late antique desert ascetics by bringing them safely within monastery walls; wandering, half-starved hermits needed to be ruled by an order. Many monastic leaders even recognized the importance of food rituals in socializing their orders' members. John Cassian's fifth-century monastic rule, one of the most important in Western monasticism, encouraged communal meals as part of the monks' daily duties.[32] The sixth-century Rule of S. Benedict also required monastic houses to extend hospitality—including meals and shelter—to travelers and guests.[33] Royal monasteries, of course, could afford the best food items both for themselves as well as their elite visitors. According to one eighth-century register, for example, the monastery at Corbie received an allowance of various spices including pepper and cinnamon from King Chilperic II.[34]

Along with royal courts, monasteries provided the primary market for the dwindling spice trade in the West. Ecclesiastic authorities relied upon food and feasts to demonstrate their charity, thereby gaining popular support, and solidifying their relationships with both elite and royalty.[35] Even with access to various spices, however, the palette of early medieval Christians certainly differed from the spice-rich cuisine of the Middle East. Medieval Europeans did not display an interest in most exotic flavorings again until after the Crusades, when pilgrims made their way back from the Holy Land with new recipes and culinary anecdotes.[36]

CUISINE IN THE ARABIAN MILIEU

Food and especially spices signify power in much the same ways in the Arabian context. Scant information about diet and cuisine survives from the pre-Islamic world, but we do have some general understandings. Most Arabs at the beginning of the Islamic era enjoyed common staples. Cereals and grains, on the whole, proved much less important than in Roman culture because of the climate and the terrain. For most, dates and camel's milk rated high above cereals. The Arabian Peninsula hosted the Bedouin, who usually traded with oasis communities for cereals; areas able to sustain farming in the Hijaz (western central Arabia) and Yemen cultivated barley along with some wheat, sorghum, millet, and sometimes rice.[37] Townspeople thus had the best access to bread; the Bedouin used grains more in gruel or stews than actually baking it. Most importantly, Arabs shared bread in customary hospitality displays: it demonstrated the giver's generosity and welcomed the guest.

As little class stratification existed in tribal groups, social status did not link necessarily with breads and grains. Women usually ground the grain, kneaded the dough, and cooked it communally; independent bakeries would have been rare.[38] Hadith traditions mention grinding grains and preparing bread in connection to the Prophet's own family. Styling the Prophet as a humble servant of God who would gain riches only in Paradise, reporters described how he mortgaged his armor to buy barley, a basic provision for his family. His followers also presented him with bread on occasions, recognizing that his household lacked even the basic staple.[39] One Shi`ite tradition even described Fatima, grinding barley for her family so diligently that it caused her hand to tear and bleed. God sent an angel to turn the wheel so the Prophet's daughter could rest.[40] The Shi`ite transmission, of course, not only highlighted the relative poverty of the Prophet's family but also his daughter's extraordinary piety.

With the Islamic Empire's rise and expansion, cereals and grains became more readily available as agricultural production grew and trade increased. Many Arabs attempted to incorporate exotic cuisines and imitate the caliphal court's high culture. This completely transformed the presence of and taste for bread (one example among many others). Those who could afford it sought out the best flour, utilized several different utensils in preparation, and then brushed the finest oils (or rosewater) on the tops. Laborers accessed the heartier (and cheaper) bread made from coarse ground wheat flour. This Arabian culture—one recently stratified socially, economically, and politically as compared to earlier Arab tribal society—evidenced a much more complex diet than what existed at Muhammad's birth in the sixth-century. Extant cooking manuals, the oldest from the tenth century, reveal much about this new, more varied Arabian palette.

Early Islamic communities (centered especially around the caliph's court) created a plethora of cookbooks, more than survive from Rome and early Christianity. Many focused on particular food types, circulated and emended over time. For example, compilations devoted only to sweets and aromatics proliferated alongside the more complete cooking guides.[41]

Three of particular interest include Ibn Sayyar al-Warraq's tenth-century *Cookery Book* (*Kitab al-tabikh*); Muhammad ibn al-Hasan al-Baghdadi's thirteenth-century *Cookery Book* (*Kitab al-tabikh*); and an anonymous (and encyclopedic) thirteenth-century Egyptian text *The Treasure of Useful Advice for the Composition of a Varied Table* (*Kanz al-fawa'id fi tanwi` al-mawa'id*). All three texts emphasize the aesthetics of eating—food preparation should include appealing presentation and color as well as scent. The later texts, while they encouraged cooks to arrange and even dye food for visual appeal, stress aroma above everything else.[42] Apicius' guide to Roman cuisine certainly encouraged dramatic presentation; Apicius delights in his guests' surprise in biting into one food and finding quite another. The Arabic cooking manuals, however, respect food for pleasure's sake only.

Al-Baghdadi begins his tome by dividing pleasures into six classes: food, drink, clothes, sex, scent, and sound.[43] He then ranks food as the most pleasurable of all these. Eating wholesome food—that is, food allowed by God—could be as enjoyable and sensual as wearing fine silks, having sex, and hearing beautiful music. The Arabic guides did not aim to shock their diners but gratify them, satisfying all their senses. Thus a truly pleasurable meal depended upon the foods' arrangement, color, and scent. For color, cooks dyed foods with additives such as saffron or arranged vivid garnishes such as eggs or parsley. Aromas were the most rapturous: coated on cooking dishes, added for flavor, and sprinkled at the end. Both al-Baghdadi and the author of *The Treasure* rely most heavily upon rosewater as a final ingredient: sprayed onto food after cooking.

Arabic cookbooks emphasize more than just the food's pleasure, of course. As their Latin counterparts, they contain information about health and dietetics, especially *The Treasure*. Borrowing from Galen's medical theory that required balance for optimum fitness, recipes often listed ingredients and their effects on the four humors—blood (hot/moist), black (cold/dry), yellow (hot/dry) bile, and phlegm (cold/wet). Unlike their Roman counterparts, cookbooks from the early Islamic Empire explain the foods' preparation in much greater detail as well as provide more variations on each dish. These versions allowed for disparate budgets (especially among cuts of meat and quantity of spices) but also for the diners' specific needs. The gifted cook met guests' specific requirements depending upon their health, mood, age, and even the season. According to the anonymous *Treasure*, for example, camphor should be added to food and drinks in hot weather and musk in cold.[44] Chefs also matched the dish's temperature to the diner's temperaments. The humoral theory ranked spices along a four point scale, with most spices listed as hot and mostly dry. Some distinctions existed, of course; pepper ranked 4 and 4 (most hot and most dry) while cinnamon ranked 2 (hot) and 1 (dry).[45]

Al-Warraq also catalogs the different humoral and medicinal qualities of spices and seasonings (*abazir*). He designated salt as hot and dry, and explained that if eaten in excess, decreased sperm. Cassia is hot and dry; it also aids in digestion and decreases gas.[46] One of the final chapters describe tasty electuaries that offer healing properties; these formulas promised to alleviate gas, stimulate coitus (warning, "not to be given to women"), and remedy colds and body aches.[47] Notably, al-Warraq explained how to make wine even though Islam forbids it; some spiced wines, he rationalized, limit the alcohol's intoxicating effects.

Arab cooking manuals had a slightly different audience than Roman guides; foods popular for elite diners served a larger percentage of the population than Apicius' *Cookery Book*. Early medieval Islam proliferated through large urban areas while early Christianity's Roman West gradually transitioned to a more rural landscape. After the Empire's dissolution, urban elites and a burgeoning court culture certainly retained some access to luxury food items and spices; however, this constituted a comparatively small group.[48] A cooking manual from Baghdad even explains that the most important difference between the Caliph's kitchens and "everyone else's" was basically cleanliness—cooks should properly clean the food and all cooking implements (and, here, cleaning cooking implements meant scenting them).[49] The manual assumes that most readers had access to the same range of ingredients as the elite.

Al-Warraq provides a list of basic ingredients for every cook's kitchen that would seem profoundly flamboyant in early medieval Gaul. He breaks down his grocery list into categories such as fresh fruits, condiments, and vegetables. Spices that should be used to scent cooking utensils form the first (and separate) group; these include musk, ambergris, rosewater, saffron, cassia, spinkenard, cloves, nutmeg, cardamom, and mastic. Essential cooking spices are peppercorns, coriander, cumin, caraway, ginger, resin (and leaves) of asafoetida, and salt.[50]

Unlike "everyone else's" kitchen art, court cooks devoted a great deal of time to food preparation. In both al-Warraq and al-Baghdadi's manuals, most of the recipes required complex practices that did not significantly alter the dishes' taste, including spicing at separate times between boils. The ability to devote both time and resources in dramatic food performances separated the caliphal court's grandeur from more mundane kitchens.

Because of well-maintained urban centers and trade routes, food crops (and their preparation styles) from India, Persia, and Africa also diffused readily.[51] This contributed to a more varied Islamic cuisine than that available in the Christian West. Eggplant, for example, originated in India yet quickly became a favorite throughout the Islamic Empire (even into Spain). Cooking manuals credit Ibn al-Madhi, the `Abbasid Caliph Harun al-Rashid's half-brother, with several recipes for dressed eggplant.[52] Its popularity clearly flourishes between the time of al-Warraq's cookbook—relatively silent about eggplant—and the anonymous *Treasure* some three hundred years later. As might be expected, accessibility to a highly differentiated cuisine signaled caliphal power and status (in this case, the `Abbasids).

Food preparation and consumption also did not occur privately; thus al-Warraq provides diners with etiquette guides and recipes for appropriate hand soaps and toothpastes. Hosts expected their guests to display a level of cleanliness and good hygiene before and after eating.[53] Mealtimes effectively displayed power and loyalty networks; guests exhibited their hosts proper courtesy. Here, al-Warraq makes clear that the boon companion (*nadim*) as well as anyone "who attends assemblies of noblemen and dignitaries" is responsible for proper manners and attire.[54] Scent plays an important role not only in preparing and presenting the food but also guest decorum. According to chapter 130, the guest's skin should radiate fragrance, and he should:

> perfume himself with incense, musk, compound perfumes, and all kinds of perfumed powders sprinkled on hair and clothes. Incense is used to fumigate

> the clothes, musk and camphor are for the hair, and perfumed powders are for the body.[55]

Diners should also be meticulous in washing their hands before meals, taking care not to touch their hair or beards before attendants serve their food. Finally, after the meal, al-Warraq directs guests to thoroughly wash their hands and mouths— away from any powerful patron's sight—and then rinse their faces and hands with rosewater. When dining with relative equals or friends, however, hand washing can be done together, with the patron washing last.

While certainly describing the latest exotic *haute cuisine* and table etiquette, Al-Warraq also details in his manual traditional Arab dishes which memorialize the foundations of Islamic identity while still transforming them into proper displays of culinary fashion and elite taste. One example, *tharid,* is a simple dish made from meat (probably lamb) and crumbled bread. The Prophet Muhammad compares the dish to his beloved wife `A'isha; it is the chief among foods as she is the chief among women.[56] Al-Warraq includes several variations of the dish, embellishing it by adding poultry and an elaborate mixture of herbs and spices. In this way, the cook commemorates past virtues and the Prophet Muhammad himself by imitating a basic culinary creation. As Roman moralists praised Rome's agricultural roots and its simple staples, al-Warraq celebrates Islam's "golden age" along with revisions that fit the tastes of the caliphal court.

Variations of *tharid* exist throughout the Islamic world, all containing the meat-bread base. Blogger "Fearless Kitchen" provides a simple version below, adapted from al-Warraq's *Cookery Book,* without a lot of herbs/spices. She includes helpful measurements and contemporary substitutions. (www.fearlesskitchen.com)

Lamb and Bread Stew:

1 pound fatty lamb, cut into pieces
1 15-ounce can chickpeas, drained
1 whole onion, peeled
1 red onion, peeled
kosher salt to taste
water to cover the meat
1 stick butter, melted
1 teaspoon sugar (approx.)
8 slices whole wheat bread—a day old or so is fine (although al-Warraq provides an easy "water bread" recipe)

1. Combine the meat, onions, and chickpeas in the saucepan. Add water and kosher salt to taste.

2. Bring to a boil, then reduce heat and cook until the meat is done.

3. Lay the bread out in your serving bowl.

4. Pour the butter over the bread. The original says to do this while the bread is still piping hot, so if you do not have fresh hot bread toast it first.

5. Sprinkle the sugar over the bread.

6. Ladle the hot stew over the bread, broth and all.

7. Serve.

Al-Warraq's table descriptions also construct elite/Muslim identity against the "other." He includes various narratives where Bedouin (*Arabi badawi*)—Arabian nomadic peoples—dine with Caliphs and other important `Abbasids, displaying their lack of civility through their poor table manners. Unlike their urban counterparts, the Bedouin failed to clean their hands, disregarded appropriate eating utensils, and were prone to gluttony.[57] Over-indulgence, like dirty hands, signaled a questionable character. Bedouin as well as other "commoners" often mistakenly ate too much, too quickly, and without concern for cleanliness. This tension contrasts earlier texts where Arabs openly ridiculed nontraditional (especially Persian) foods; and, Arabian tribal "strongmen" scoffed at the luxurious ways of the Persian table. Al-Warraq's cookbook displays just how varied (and elitist) Islamic cuisine—and identity—had become. *The Treasure*, compiled much later, includes the most wide-ranging recipe collection, linked to Damascus, Baghdad, Spain, the Yemen, and Egypt.

Food rituals throughout the Roman and Arabian worlds incorporated pleasures beyond taste to include sight (beautiful presentation) and smell (from both ingredients and added fragrance). Arabic food manuals provide an important glimpse into early Islam's social, political, and cultural diversification, especially during the caliphates. The recipes drew upon ingredients from across the growing empire, signaling its power and its ethnic diversity (as compared especially to its Arabian roots). Within the empire, most cooks regardless of status could still acquire several spice varieties. Once again we see a stark contrast with the Roman world as access to spices and perfumes became increasingly confined to the Church and ruling elite.

Cooking manuals also suggest something more basic about Islamic culture—its early obsession with physical purity. All three compilations discussed here demonstrate a profound emphasis on cleanliness and scent; for example, cooks should use fragrance on meat *and* their body; carefully trim their nails; and closely examine pots and pans for dirt or contagion. Al-Baghdadi's high expectations for food preparation complement Islam's general standards of ritual purity.[58]

According to Islam, Allah requires physical purification before various ritual actions (such as prayer), obligating Muslims to be aware of their physical state throughout the day and night. Just as Jewish culinary rituals required separating themselves from "impure" Gentiles, the Islamic inclusion of physical purity before, during, and after meals (and their preparation) confirms their own distinctiveness. Such food rituals thus successfully constructed religious and ethnic identity—with concomitant displays of status and power—against the outside "other." Food protocols, especially those in Islam, idealized physical purity and even equated it with spiritual health. Yet, Arabs and Romans alike entwined physical well-being with God's grace; disease with God's judgment or personal fault. Disease threatened the individual body as well as the body social; it could attack when least expected from a variety of forces. We now turn to this topic, focusing on scent, healthcare, and disease.

NOTES

1. Al-Warrāq 3, p. 92.
2. Apicius, *Apicius: A Critical Edition With An Introduction And An English Translation Of The Latin Recipe Text*, trans. Christopher Grocock and Sally Granger (Totnes, UK: Prospect Books, 2006), 1.13.
3. Al-Warrāq, chapters 8–25 (with some exceptions therein).
4. There are exceptions to this generalization. Some Romans were vegetarian, for example, and some eschewed beans. Michael Beer presents a fine discussion of voluntary food restrictions in *Taste or Taboo: Dietary Choices in Antiquity* (Devon: Prospect Books, 2010).
5. See Jordan D. Rosenblum's *Food and Identity in Early Rabbinic Judaism* (Cambridge, UK: Cambridge University Press, 2010), 17.
6. Rosenblum, 56.
7. *Jubilees* 22.16.
8. *T Avodah Zarah* 4.6 (concerned particularly for wedding feasts) and *T Hullin* 1.1; discussed in Rosenblum, 75–101. As Rosenblum points out, however, the tannaitic texts also include a variety of leniencies to allow for the requisite interaction between Jews and non-Jews. David M. Freidenreich does an especially fine job of distinguishing the subtle differences between Mishnah and Talmud regarding separation of Jews and Gentiles. He argues that Talmudic restrictions allow for gentiles (for example) to labor with Jewish food preparers during some stages of food preparation—highlighting religious identity without forbidding interaction with gentile culture. See *Foreigners and Their Food: Constructing Otherness in Jewish, Christian, and Islamic Law* (Berkeley: University of California Press, 2011).

9. *T Hullin* 8.1–3. See Rosenblum, 140–43. Rosenblum also argues in his work that these dietary laws also serve to distinguish rabbinic Jews from non-rabbinic Jews.

10. See the introductory material by Christopher Grocock and Sally Grainger in their translation, *Apicius: A Critical Edition with an Introduction and an English Translation of the Latin Recipe Text*, 13–38.

11. Asafoetida, also called "Devil's Dung" is a foul-smelling resin taken from a tuber plant; it is generally sold in either block or powdered form. Although its initial aroma is quite disgusting, it becomes pleasant and succulent when heated (for example, in olive oil). Greek herbalist Dioscorides taught that asafoetida served as a general cure all for everything from goiters to baldness, to lung disease. Chip Rossetti discusses his recent encounters with the aromatic in "Devil's Dung: The World's Smelliest Spice," *Saudi Aramco World*, July/August (2009): 36–43.

12. Scholars generally disagree about exactly what differentiates the two. *Garum* is produced, basically, by allowing salt and fish to ferment. Some suggest that consistency of the sauce (from very liquid to almost a paste) is what distinguishes *garum* and *linguamen*. Others suggest the two terms refer to the same substance.

13. Petronius, §33, 36.

14. Apicius, *Apicius: A Critical Edition With An Introduction And An English Translation Of The Latin Recipe Text*, 4.2.14.

15. See Beer, 20–21.

16. The law also limited the number of guests at dinner parties and exempted domestic produce from price limitations thus discouraging a market dependent upon imported goods. See Vincent J. Rosivach's review of the law in "Lex Fannia Sumptuaria of 161 BC," *The Classical Journal* 102.1 (October–November 2006): 1–15. Also, John H. D'Arms discusses the *Lex Fannia* in "The Culinary Reality of Roman Upper-Class *Convivia*: Integrating Texts and Images," *Comparative Studies in Society and History*, 46.3 July (2004): 428–50.

17. John H. D'Arms proposes this important and convincing argument in "The Culinary Reality," 434–50.

18. K. Dunbabin, "Triclinium and Stibadium," in *Dining in a Classical Context*, ed. William J. Slater (Ann Arbor: University of Michigan Press, 1991), 121–48.

19. See Grant's introduction to his translation, *On the Observance of Food*, 26–28.

20. Anthimus, intro., p. 49.

21. Anthimus repeatedly warns against raw meat especially; see among others 4, 23, pp. 53, 59.

22. Anthimus 17–18, p. 57.

23. See Anthimus 3, p. 51, his "spiciest" recipe.

24. Anthimus 15, p. 57.

25. Anthimus 14, pp. 55, 57.

26. Anthimus 76, p. 77.

27. Anthimus 80, p. 79.

28. See Mintz and DuBois' discussion of current approaches to food rituals in "The Anthropology of Food and Eating," *Annual Review of Anthropology* 31 (2002): 99–119.

29. See Bonnie Effros' discussion in *Creating Community with Food and Drink in Merovingian Gaul* (New York: Palgrave Macmillan, 2002), 64–66.

30. See E. Dench discussion in "Austerity, Excess, Success, and Failure in Hellenistic and Early Imperial Italy," in *Parchments of Gender* (Oxford: Oxford University Press, 1998), 121–46.

31. See Grimm's fine discussion of Jerome's ascetic agenda, 148–68.

32. Cassian, *De institutis coenobiorum* 1.11 and 3.12.

33. Benedict of Nursia, *Regula*, 5.

34. Discussed in Effros 30. See Léon Levillain, *Examen critique des chartes mérovingiennes et carolingiennes de l'abbaye de Corbie* 15. Mémoires et documents publiés par la Société de l'École des chartes 5 (Paris: A. Picard et Fils, 1902).

35. See an introduction to this topic in Danuta Shanzer's "Bishops, Letters, Fast, Food, and Feast in Later Roman Gaul," in *Society and Culture in Late Antique Gaul: Revisiting the Sources*, eds. Ralph W. Mathisen and Danuta Shanzer (Burlington: Ashgate, 2001), 217–36.

36. Toby Anderson also suggests an interest in Arab cuisine because of its role in paradisiacal religious literature in Islam. See "The Arab Influence on Western European Cooking," *Journal of Medieval History* 6 (1980): 317–40. Nawal Nasrullah argues such European imitation of Arab cuisine was due to the reconquest of Spain and Sicily; al-Warrāq 29.

37. See David Waines' discussion in "Cereals, Bread and Society: An Essay on the Staff of Life in Medieval Iraq," *Journal of the Economic and Social History of the Orient* 30.3 (1987): 255–85.

38. See Waines "Cereals, Bread and Society," 269.

39. Bukharī 48.1.

40. Al-Majlisī 43.3, pp. 28–29.

41. See Marin and Waines discussion of the dissemination of various cookery books in their edition of *Kanz*, 5.

42. See a good introduction to the aesthetics of Arab foods in Manuela Marin's "Beyond Taste: The Complements of Colour and Smell in the Medieval Arab Culinary Tradition," in *Culinary Cultures of the Middle East*, eds. Sami Zubaida and Richard Tapper (New York: I. B. Tauris Publishers, 1994), 205–14.

43. Al-Baghdadī, 37.

44. *Kanz*, 12, 33, 49, among many other recipes.

45. See Nawal Nasrallah's discussion in introduction, 61.

46. Al-Warrāq 18, pp. 136–37.

47. Al-Warrāq, chapter 125, pp. 481–87.

48. David Waines discusses this culinary sophistication in "'Luxury foods' in Medieval Islam Societies," *World Archaeology* 34.3 (2003): 571–80.

49. Al-Warrāq 1, p. 81.

50. Al-Warrāq 3, pp. 91–93.

51. Relating more to the actual food items that spread throughout the early Islamic Empire, see Andrew Watson's "The Arab Agricultural Revolution and Its Diffusion, 700–1100," *Journal of Economic History* 34.1 (1974): 8–35.

52. Al-Warrāq 45, pp. 226–27.

53. Al-Warrāq 19, pp. 495–502.

54. Al-Warrāq 130, pp. 502–3.

55. Al-Warrāq 130, pp. 502–3.

56. Bukharī, v. 7, p. 262; discussed in Waines, "Cereals, Bread and Society," 266.

57. Al-Warrāq 131, pp. 508–10.

58. Al-Baghdadī, 37–39.

Chapter Three

The Smell of Health and Disease

Both Romans and Arabs associated disease and death with physical and spiritual impurity, an impurity that threatened social boundaries as well as the eternal soul. This worldview likened sickness to foul smells and health with pleasing fragrance.[1] Several hadith, for example, explained that musk rectified "bad air" and protected against disease.[2] Sweet aromas, either burned as fumigants or worn as pomanders, often presented the only recourse against the spread of sickness and plague. Fragrance—guarding against impurity and bad smells—thus functioned as an apotropaic defense against illness as well as a curative agent that could replace the decay of infection with the scent of wellbeing. Extant medical manuals include various prescriptions that promote sweet smelling spices and perfumes' healing properties.

The Romans absorbed much of their medical knowledge from the Greeks, based mainly on the Hippocratic corpus, even though this caused some controversy during the late Republic.[3] For statesmen like Cato, the Roman *paterfamilias* should be ultimately responsible for his estate's health, including his family, slaves, and even animals. He decried foreign, exotic influences on Roman identity, including the Hellenization of healing. Family leaders should be acquainted with proven folk remedies and rituals that secure health and healing; for example, Cato includes one prescription for intestinal and stomach worms comprised of pomegranate blossoms, frankincense, wine, honey, and jumping from a square pillar ten times.[4] By the mid-first-century BCE, health professionals and Greek medical practices proliferated widely even though they had been stigmatized as "foreign" or non-Roman. Celsus, too, commented on healing practices in *On Medicine*, at once congratulating the medicinal prowess learned from the Greeks but labeling its professionalization as "foreign," threatening Roman cultural superiority. For him, cures could be learned by anyone and should never be dispensed for profit.[5]

Among these late Republican and early Imperial suspicions, a vigorous Roman medical tradition developed nonetheless, built both in-line-with and in-opposition-to its Hippocratic ancestors. Romans themselves boasted several important physicians including Asclepiades, Scribonius Largus, and Dioscorides; yet, the most important figure is Galen (d. c. 217), recognized as one of the most pivotal figures in Western medicine.[6] One of Galen's basic approaches to the body and health included dietetics—regulating food and drink to both cure and maintain good health. All foods and spices, as mentioned in the previous chapter, could be categorized along a spectrum of the four humors and their qualities. If dietetics failed, only then should medicine be prescribed, also according to humoral theory.

Galen viewed a healthy diet as one filled with diverse foods, complementing each other in their humoral qualities; yet, ultimately, a diet's restorative effect depended on the individual's own nature. Galen surveys most Mediterranean staples in his text, *On the Powers of Food* (*De alimentorum facultatibus*). He divides the work into three sections focusing on cereals, grains, plants, and animals. He notes, for example, that lentils can be prepared in numerous ways that serves as a laxative and, in contrast, "dry fluxes in the stomach." The laxative effect would benefit an individual of "watery disposition" yet detrimental for those already "too dry."[7] Various spices and fragrance could also enhance humoral qualities in most recipes. Thus food—and their seasonings—should be chosen carefully; a physician operated as much as a cook as anything else.

Galen discusses his views on food and health in "Thinning Diet" (*Subtiliante diaeta*); herein, he notes that most chronic diseases can be treated through diet instead of pharmaceuticals.[8] The most effective regimen "thins" the humors; the senses of smell and taste indicate what food types lead to thinning. Most importantly, foods that smell irritating, striking, even unpleasant, suggest they will provide a cutting or thinning process in the body. Garlic, onion, and leeks function in this way. He then goes on to list basic vegetables, herbs, fruits, meats, and drinks that combat thick, sticky, and phlegmatic humors related to illness. Wine vinegar mixed with honey imparts the greatest healing qualities and even serves as a base for many other medicines. For Galen, everyone has access to remedial foods, yet only the skilled healer fully appreciates their qualities and understands how to diagnose illness properly. The best among physicians can indeed take one curative, one ingredient, and use it to treat multiple disease; and, only the elite can know that when mixed with different media, such as vinegar, oil, or rosewater, medicines act differently. Only those dedicated to the medical arts fully discern food's subtle properties and how to apply them in healing.[9]

This basic premise that a "balanced diet" leads to good health related to more than just the physical body. From a more philosophical perspective, akin to Stoic teachings, good health and dietary habits reveal much about a person's character. Most individuals can control their dietary intake; to make poor, unhealthy, or ill-informed choices betrays moral deficiencies. For Galen and other like-minded physicians, moderate and balanced food choices signaled wisdom and self-discipline; anything less led to one's enslavement to gluttony and desire.[10] Roman moralists criticized these entrapments and viewed them as a threat to the body social: who could trust political leaders and citizens who lacked the basic virtues of self-control?

In the ancient world, good health and happiness indicated not only moral character but also divine favor. Healers never divided their art between "secular" and "religious"; instead, the ancient world observed multiple etiologies of disease. Even physicians dedicated to empirical observations of nature often noted that god(s) created the material world and thus they could confidently trace disease to material causes. This approach did not rule out the possibility of sickness as a form of divine punishment or demonic possession (*daimones*) and healing as the manifestation of divine intervention. In classical Rome, the cults of Asclepius, Isis, and Serapis, in particular, addressed concerns for health and healing.

Most ancients saw no problem in praying and making propitiations to the god(s) when ill, while at the same time consulting a physician for remedies. The god(s) could miraculously cure as they willed or imbue holy men and women with healing charisma; they could just as easily direct the patient to a healer, potion, or exercise regimen. Asclepius often divulged plant and mineral-based curatives or various other ways of manipulating the physical body though physical means. Within the same worldview, then, healing could be traced to natural means or miraculous intervention, or both at the same time.

These complementary etiologies continue in early Christianity, just as it had in Rabbinic Judaism. According to the Hebrew Bible, disease could be interpreted as God's punishment (Exodus 12:12; 1 Samuel 5:6; 2 Chronicles 26:20); yet this did not negate physicians' skill or the medical arts (2 Kings 20:7). Hellenized Judaism assimilated many of the Greco-Roman medical traditions, including the healer's professional status. According to the Apocryphal Book of Sirach:

> Honor physicians for their services, for the Lord created them; for their gift of healing comes from the Most High, and they are rewarded by the king. The skill of physicians makes them distinguished, and in the presence of the great they are admired. The Lord created medicines out of the earth, and the sensible will not despise them. And he gave skill to human beings that he might be glorified in his marvelous works. By them the physician heals and takes away pain; the pharmacist makes a mixture from them. God's works will never be

> finished; and from him health spreads over all the earth. My child, when you are ill, do not delay, but pray to the Lord, and he will heal you. Give up your faults and direct your hands rightly, and cleanse your heart from all sin. Offer a sweet-smelling sacrifice, and a memorial portion of choice flour, and pour oil on your offering, as much as you can afford. Then give the physician his place, for the Lord created him; do not let him leave you, for you need him. There may come a time when recovery lies in the hands of physicians, for they too pray to the Lord that he will grant them success in diagnosis and in healing, for the sake of preserving life. He who sins against his Maker will be defiant towards the physician. (38:1–4, 6–15)

For Jesus ben Sira, healers gain their gifts from God and should be consulted—after the sick first pray to God and make sweet-smelling offerings for their sins.

Natural and miraculous medicine occur in the New Testament as well. Demons cause both mental and physical anguish, and Jesus casts them out (Mark 1:23–26; Luke 4:33–35). Jesus also corrects various congenital illnesses including deafness, blindness, leprosy, and paralysis (see Mark 7:31–37, 8:22–26; Luke 17:11–19; Matthew 9:28). Confronting traditional Jewish purity rituals, He heals a woman's issue of blood after she touches His robes "with faith." (Mark 5:25–35) And in his epistle, Paul directs Timothy to "take a little wine" for his stomach ailment, a cure Galen would have approved (1 Timothy 5:23).

Early Christians identified Jesus as the Great Physician, the healer of sinners (Mark 2:17; Matthew 9:12; Luke 5:31), which suggests a cure of soul as well as body. Church historian Eusebius (d. c. 340 CE) even quotes from the Hippocratic work *On Breaths* in describing Jesus: "A devoted physician . . . in treating another man's troubles brings suffering on himself."[11] The Gospels even credit Jesus' apostles with miraculous healings which prove the truth they preach. The presence of charismatic cures, however, did not necessitate the denial or avoidance of natural remedies.

Early Church Father Origen (d. 254 CE) recognized God as the supreme healer who mercifully provides medical knowledge to human beings. God thus gifts medicine and physicians to humanity who may then righteously consult them.[12] Origen introduced a dichotomy that continued into the early Middle Ages—very pious Christians, such as later ascetics, relied upon only God as Healer while quite rightly interpreting their illness as a punishment or simply God's will.[13] But those folks formed the minority; the majority of Christians saw no problem with consulting physicians, as long as they understood that medical intervention worked only if God willed it. Translations of important medical guides such as Dioscorides' *Materials of Medicine* continued into the early medieval period; and most early medieval courts and monasteries maintained a staff of doctors.[14] Gregory of Tours recognized that although seeking physicians' care sometime demonstrated a Christian's lack of faith, it was also wise and appropriate because God had allowed them their skills in the first place.[15] Indeed, Gregory probably read some medical works as he includes at least twenty-eight technical medical terms, mostly from Greek, throughout his works.[16]

While various etiologies and remedies might have existed in early medieval Gaul, the most celebrated form of healing remained the Church and its saints.[17] They offered healing that was more than just skin deep; it remedied the soul as well. Gregory of Tours called S. Martin the "Great Doctor" as he encouraged his veneration. He compared the saint's healing with that of a regular doctor:

> Oh indescribable antidote! O unspeakable balm! O praiseworthy remedy! O heavenly purgative, if I may say so! This dust [of S. Martin] overwhelms the subtleties of doctors, surpasses sweet scents, and is more powerful than all strong ointments. . . . Not only does it strengthen disabled limbs but—something that is more important than all these—it removes and lightens those very blemishes of conscience.[18]

Gregory does not herein dismiss the efficacy of physicians and medication; he only praises the superiority of saintly remedies.

Early medieval hagiography listed precise recipes for healing unguents obtained usually at cultic sites. Instead of relying upon only herbs and pharmaceuticals, the pious could also resort to the saints' physical remains in various forms. Gregory of Tours' hagiographies are replete with recipes for mixing ash and water, oil, and even candle wax to bring about miraculous cures. For example, candle wax taken from S. Martin's tomb cured a "deaf and dumb" woman believed to be possessed by a demon after she placed it in her ear.[19]

Late antique and early Christians' varied etiologies contextualized illness and suffering in slightly different ways than their pagan and Jewish counterparts. Physical suffering certainly could be punishment from God—as most ancients also accepted—but it also extended a unique path to salvation itself. Suffering was redemptive; it linked the Christian to the body of Christ, which had suffered unimaginable pains and torments (discussed more fully in the next chapter). In this scenario, the disease, the sickness, the physical ailment became sweet; it purged the body of sin and transformed the suffering into salvation. The disease was more aromatic than the cure.

MEDICINE IN THE ARABIAN MILIEU

Arab medical knowledge also addressed health and healing with various etiologies; however, the Muslim world never relied so heavily upon the miraculous, saintly paradigms when confronting disease. Galen's humoral theory reigned supreme in most physicians' guidebooks; yet, Muslim authors always recognized divine will or to the (at times malevolent) jinn, supernatural creatures that often taunt or trick humans. The varied etiologies available to early Muslims appear in collections of medical advice attributed to the Prophet Muhammad.[20] Scholars such as Shams al-Din al-Dhahabi (d. 1348) and Jalal al-Din Suyuti (d. 1505) arranged prophetic hadith relating to healing and medicine with their own commentary, philosophy, and legal spins (al-Dhahabi, for example, followed the Hanbali legal school). Other collections of *Tibb al-nabi* (or, *Medicine of the Prophet*) probably date back to the ninth-century.[21]

Prophetic hadith and their commentaries provide a glimpse into the early negotiation that occurred between Arabian folk custom, Islam's new teachings, and the medical traditions assimilated into Islamic culture largely through Greek, Persian, and Indian texts. The prophet, for example, acknowledged that both disease and healing come only from God; unbelievers see plague as Divine punishment while Muslims rejoice in it as God's will.[22] According to Islamic tradition, the prophets themselves—God's chosen and beloved—experienced illness thus it should not be interpreted as God's wrath.[23] One tradition noted that "there is no disease that Allah has created, except that He also has created its treatment."[24] Physicians used such hadith to argue for medical sciences which depended heavily upon Greek and Persian "foreign influence": if God provided cures, then physicians should set their mind to understanding them. This did not preclude Allah's direct intervention; sometimes miraculous healing occurred. The Prophet Muhammad occasionally affects God's healing, but the miracles are much less spectacular than those of Jesus reported in the New Testament. One remedy, for example, required rubbing a wound with a believer's saliva mixed with soil while reciting Qur'an passages (especially the opening chapter, *Surat al-fatiha*).[25]

More than anything else, the Prophet offered healing advice. He emphasized two particular cures: drinking honey for various conditions and undergoing cupping (a procedure which attracts blood to the surface).[26] Some of his medicinal suggestions coincided with curatives described in other medical manuals that drew upon local aromatics. Incense, ground and sniffed, cures pleurisy; black cumin, crushed and inserted in the nostrils, relieves "all diseases except death."[27] Other advice resembled local folklore and Arab custom, attributing some maladies to jinn, the presence of the evil eye, or magic. The Prophet proposed cures for these disorders as well. Eating several dates each morning protected against poison or magic; wearing certain amulets deflects the evil eye.[28]

The Prophet himself once realized that he suffered from false memories because of an enemy's spell; his adversary had conjured a spell by placing his comb and hair within a date skin by a well. He located the charm but noted that Allah had eradicated its power.[29] In this instance, Muhammad denounced the enchantment as evil paganism. Yet, the Prophet reportedly approved of amulets for believers' protection against disease and curses; such amulets should include the words of the Qu'ran, recognized for their healing qualities.[30] Al-Suyuti rehearses various scholars' arguments that as long as believers understood the amulet's words and source of healing (i.e., God), they were permissible.[31] That is the compromise: Arab customs relating to causes and cures of disease (as well as bad luck) continue only within the larger paradigm of God's omniscience and omnipotence.

The paradigm of spiritual healing, either directly through Allah or more "magical" forms, did not proliferate as aggressively as it did in early Christianity. This may be because, in the West, medical knowledge and (eventually) hospital care came to reside in monasteries, churches, or ecclesiastical schools. Christian theology largely directed medical theory and practices. Physicians in the Islamic world—certainly working within a religious context, ultimately attributing all healing to Allah—experienced a more secular divide. Governmental and private endowments supported medical training based in Greco-Roman tradition as well as the establishment of hospitals where non-Muslims often operated alongside Muslim colleagues. In such a context also without a centralized, charismatic clergy empowered with healing miracles, suffering, and disease did not become associated with sin and redemption. In the Islamic world, most physicians continued to gauge disease according to Galenic models in addition to Indian and Persian techniques while the Christian West turned increasingly to a more spiritual pathology.[32]

There is indeed a surfeit of medical manuals and guides that survive from the early Islamic world before the rise of great physicians and intellectuals such as Ibn Sina/Avicenna (d. 1037) and Ibn Rush/Averroes (d. 1198). For our purposes, we will examine three of these. Perhaps the earliest extant Arabic pharmacology is Sabur ibn Sahl's dispensatory guide. Ibn Sahl (d. 869 CE), a Nestorian Christian from Persia, served as a medical adviser in Baghdad for `Abbasid caliph al-Mutawakkil (r. 847–861 CE). Second, al-Kindi (d. 865) wrote a medical formulary (or, *Aqrabadhin*). He has a notable interest in aromatics; he (or one of his students) also compiled the *Book on the Chemistry of Perfume and Distillation* (or, *Kitab fi-kimiya' al-`itr wa l-tas`idat*). This handbook contained several recipes for *sukk* (medical compounds of various aromatics). Finally, an extant anonymous *Syriac Book of Medicines*, part of it written in the early-Christian era in Greek and then translated into Syriac, influenced the Arab medical world. The first section is based on "science"; the second on astrology; and the third consists of "native medicines." This final section includes various folkloric recipes, popular practices that rely upon magic and animal products along with various aromatics.

Ibn Sahl's text, which existed in various editions, provides multiple curatives for a plethora of symptoms, drawing upon both Greek and Persian customs.[33] The recipes usually contain the ailment's description and then a list of ingredients—including measurements—and preparation guide. Almost all formulae reference the body's humors or humoral qualities in some way; the author also mentions Galen and (probably) Dioscorides.[34]

All three ancient texts have a few things in common. First, all three rely heavily upon aromatics in their curatives. Ibn Sahl's first chapter lists a variety of pastilles; most contain multiple ingredients including musk, myrrh, saffron, cassia, clove, and roses. Several of the pastilles have rose-water as their base; liquid medicines include spiced wines, milk, and various fruit juices.[35] Ibn Sahl employs s*ukk* in several recipes, as do other physicians in medieval Islam, such as al-Kindi. Their s*ukk* recipes include (among other items) dates, mace, clove, cardamom, sandalwood, aloeswood, and musk. Ibn Sahl recommends medicines with *sukk* to combat hot humors; al-Kindi includes them in recipes for sore throats and "sexual overindulgence."[36]

The anonymous *Book of Medicine* also incorporates a variety of aromatic remedies in its approach to the body and healing. In a section on "nervous diseases"—which relate to the body's nervous system and problems with rigidity, paralysis, and spasms—the physician prescribes unguents to anoint the head. If the patient is without fever or inflammation, oils of chamomile, narcissus, and nard are used; if fever is present, substitute with oils of roses or violets. The text offers an "Indian" and "Persian" version of the cure which includes peppercorns and camphor oil respectively.[37] The "native prescriptions" also rely upon aromatics; the author prescribes a mixture of frankincense and crocus, mixed with egg white, for a head wound.[38] Such recipes would lie within reach of most household economies as they included few and (fairly) easily accessible ingredients. As with medical guides for the poor, the text also lists popular healing folklore accessible to everyone. Some cures called for dried and powdered animal dung; one cure for poison requires imbibing children's urine mixed with wine.[39]

Second, the three authors include fairly conventional ailments and their cures, but they do not stop there. Ibn Sahl's pharmacology, for example, addresses simple health concerns as well as prescriptions for alleviating pain (what we might call today "over the counter" curatives?). He recommends one poultice to relieve venomous stings: this includes castoreum, myrrh, spikenard, cassia, saffron, and opium, soaked in a wine reduction.[40] Al-Kindi notes the use of ground myrrh as a general analgesic for toothaches.[41] The three authors even offer assistance with more aesthetic concerns: curatives for freckles; concoctions to promote clear skin; a paste for ridding oneself of lice; cures for the "hot-tempered"; dyes for the hair; and remedies that remove warts.[42]

Finally, all three texts devote considerable space to concoctions concerned for sexual pleasure and contraception.[43] Ibn Sahl includes forms of contraception for both men and women; one formula required a woman to drink dyer's indigo in water to prevent pregnancy for one month.[44] Other recipes list ingredients such as tar and sesame or olive oil that men might swab over the penis before intercourse. Ibn Sahl includes a recipe that "narrows the woman's vagina and restores her virginity"—the intention here is not to refurbish the hymen but to narrow the vagina for the male's pleasure. It calls for a cotton cloth to be soaked in iris essence, sprinkled with oak gall powder, and inserted into the vagina before intercourse.[45] Al-Kindi's recipes also address women's sexual stimulation and pleasure. In one formula, he recommends mixing oil of jasmine and asafoetida together, letting it steep, and then applying to the penis before sex. He also proposes this as a remedy for women "without desire" in general.[46] These medical manuals thus considered women's sexual stimulation and pleasure separate from procreation expectations.

Another important medieval physician, Persian al-Razi (d. c. 925), commented most extensively on contraception and abortifacients available to Muslim men and women.[47] Within Islamic tradition, sexual stimulation and satisfaction factored as part of God's gifts to humanity; sexual intercourse may lead simply to pleasure instead of conception. Thus Arab physicians discussed various methods of preventing or ending unwanted pregnancy.[48] Al-Razi in particular drew upon a long medical tradition from Greco-Roman authors including Hippocrates, Galen, and Dioscorides. Many of the recipes mentioned in al-Razi were indeed available to women of classical Rome and early Christianity.[49] Some of the early Christian authors, perhaps avoiding theological arguments that linked conception with original sin, referred to abortifacients as methods to "induce" or "delay" menstruation.[50]

Some of the common ingredients contained in contraceptives and abortifacients across Roman, early Christian, and Islamic traditions include rue, iris, Artemisia, saffron, and pepper. Pliny even warned his readers against using rue in recipes because it could act as an abortifacient.[51] Al-Razi and others discuss different media for the medicines; some could be taken orally by the female; inserted into the vagina as a suppository or tampon; fumigation of the vagina and womb; and (occasionally) applications for the penis. One recipe for an oral abortifacient from an early Christian source requires pepper, saffron, myrrh; al-Razi mentions some simple combinations of cinnamon and myrrh or wild rue seeds.[52] Suppositories could incorporate juice of an onion; iris with honey; juice of peppermint before coitus and pepper after coitus.[53] Fumigation materials included cardamom and cyclamen.[54]

Both Roman and Arab pharmacologies agreed that spices and other aromatics affect the body: they can provide birth control, curatives, and even poisons. The same assumption held true for psychology; portending the concerns of aromatherapy, scent modulated the mind and the spirit. Certain mixtures of scent such as aloeswood, myrtle, violet, jasmine, and marjoram inspired pride, generosity, and nobility. Others promoted pleasure, yearning, and desire; this recipe called for rose, narcissus, and wallflower. Not surprisingly, perhaps, al-Kindi labeled the first combination masculine and the second as feminine.[55]

Scents play such an important role in Arab medical manuals that physician Ibn Masawaih (d. 857) compiled a pharmacology devoted solely to their healing properties. A Christian, Ibn Masawaih, served various caliphs as personal physician, trained numerous young doctors, and wrote forty-four books.[56] One text, *Simple Aromatic Substances* (*Kitab jawahir al-tib al-mufrada*), focuses first on the five "primary" scents: musk, ambergris, aloe, camphor, and saffron. Ibn Musawaih then turns to twenty-four "secondary" aromatics including nard, clove, nutmeg, pepper, cardamom, and nutmeg. Of the main (and most expensive) scents, he distinguishes between lesser and greater grades and even warns against falsified forms. Musk, he explains, benefits the heart and stops bleeding;[57] ambergris balances the humors of the aged;[58] aloe treats the nerves, liver, and stomach;[59] camphor keeps wounds from stretching;[60] and saffron addresses all the humors.[61] Along with healing properties, Ibn Musawaih discusses popular cooking combinations (e.g., clove and nutmeg) as well as applications in perfume and incense.[62] He confirms for us that Arabs valued a large variety of scents for a large variety of purposes.

CONCLUSION

In this Part I, we have reviewed some of the most important functions of scent in the Roman and Arabian milieus at the time of early Christianity and Islam. As we have seen, these religious traditions emerge in very distinct cultural milieus, with accessibility to scent declining in the Roman world yet proliferating throughout the Islamic lands. While the Church advertised its power with fragrant ceremonies, scent and spice remained within reach of most residents under the Umayyad and Abbasid caliphates. The Church also administered scented healing within its sacred spaces; in the Islamic world, hospitals flourished with governmental or private endowments.

Within both worlds, however, men and women enjoyed fragrance in multiple ways: as part of their personal toilet; as aphrodisiac as well as contraceptive; to flavor food; to soothe the sick and anoint the dead. Beyond the physical body, however, scent coded the social body as virtuous or profligate; as effeminate or virile; as pious spouse or wonton whore. The presence of scent bound together the body social, delineating community boundaries and cueing hierarchy. Spectacles in the arenas and theaters—and those rehearsed in martyr narratives and poetry—united the audience (viewers and hearers) by arousing corporal attentions and desires familiar to all. Within the body social, scent effectively identified the "other," whether that be uncivilized Bedouin or subordinate ethnicity.

Within these sensory experiences, both individual and shared, fragrance effectively marked important moments of transition in both time and space. Romans and Arabs used scent in commemorating rites of passage such as marriage and death while, at the same time, fumigating their homes and altars in distinguishing private from public, health from sickness; and sacred from profane. It is to this latter function that we now turn our attention—the diverse ways both Christians and Muslims employed aromatics in constructing and identifying the sacred.

NOTES

1. See, among many others, the discussion of Annick LeGuérer, *Scent: The Mysterious and Essential Powers of Smell* (New York: Kodansha International, 1994), 64–66.
2. Al-Suyūṭī, *al-Maqama al-miskiyya,* 113–14.
3. For a fine survey of smell in ancient healing, see Laurence Totelin's "Smell as Sign and Cure in Ancient Medicine," in *Smell and the Ancient Senses,* ed. Mark Bradley (London: Routledge, 2015), 17–29.
4. Cato, *De agricultura* 127, 1–3. Discussed in Vivian Nutton, *Ancient Medicine,* (New York: Routledge, 2004), 162.
5. Celsus, *De medicina*, especially 3, 4, 8–16; discussed in Nutton 166–67.
6. See Nutton for an important review of medicine in the Roman world. He includes, for example, important discussions of what Galen borrowed (and vigorously rejected) from his near contemporaries.
7. Galen, *De alimentorum facultatibus*, trans. in *On Food and Diet*, 96–97.
8. Galen, *De subtiliante diaeta*, 305–24.
9. Galen, *De optimo medico cognoscendo* 13.1–3, pp. 129–30.
10. For an important overview of Greco-Roman views of morality and food see Grimm, chapter 2.
11. Eusebius, *Church History* 10.4.1; discussed in Gary B. Ferngren, *Medicine and Health Care in Early Christianity* (Baltimore: Johns Hopkins University, 2009), 30–31.
12. Origen, *Contra celsum* 3.12. Discussed in Ferngren 26–27.
13. Origen, *Contra celsum* 8.60. Discussed in Ferngren 26–27.
14. See, for example, Andrew Crislip, *From Monastery to Hospital: Christian Monasticism and the Transformation of Health Care in Late Antiquity* (Ann Arbor: University of Michigan Press, 2005).
15. See Edward James' discussion in "A Sense of Wonder: Gregory of Tours, Medicine and Science," in *The Culture of Christendom,* ed. Marc Anthony Meyer (Rio Grande: Hambledon Press, 1993), 45–60.
16. See M. Bonnet, *Le Latin de Grégoire de Tours* (Paris: Librairie Hatchette, 1890), 218–20; discussed in James 55.
17. Gerhard Baader describes the dismal state of medical guidebooks in western Europe, noting the fragile translation patterns between Greek and Latin. See "Early Medieval Latin Adaptations of Byzantine Medicine in Western Europe," in *Dumbarton Oaks Papers,* v. 38, Symposium on Byzantine Medicine (1984): 251–59.
18. Gregory of Tours, *Wonders of St. Martin*, iii, 60; Krusch II, 197; trans. Peters.
19. Gregory of Tours, *Gloria confessorum*, 9–10; Krusch, 303–4.

20. Evidence of "nonscientific" approaches to medicine is also located in medical guides directed to the poor. In these manuals, authors assume limited access to the various spices and herbs required for medicinal remedies and rely upon more common items. Animal products, such as urine and feces, often function more prominently in such curatives. Some cures appear more like magic or superstition to a modern reader; e.g., hanging hare's excrement on a woman to ease parturition. See tenth-century physician Ibn al-Jazzār's *Ṭibb al-fuqarā' wa 'l-masākīn* (*Medicine for the Poor and Destitute*). This genre of medical texts also existed in the Western Christian world; indeed some scholars believe that Constantine the African translated Ibn al-Jazzār's monograph in the tenth century (entitled *Liber pauperum*). See Gerrit Bos' "Ibn al-Jazzār on Medicine for the Poor and Destitute," *Journal of the American Oriental Society* 118.3 (July–September 1998): 365–75.

21. See Cyril Elgood's discussion of various collections in his introduction to al-Suyūṭī's text, "Ṭibb-ul-Nabi or Medicine of the Prophet," *Osiris* 14 (1962): 41.

22. Bukharī 76.49.

23. M. I. H. Farooqi, *Medicinal Plants in the Traditions of Prophet Muhammad: Medicinal, Aromatic and Food Plants Mentioned in the Traditions of Prophet Muhammad* (Lucknow: Sidrah Publishers, 1998), 24.

24. Bukharī 76.1. Al-Suyūṭī also discusses God's providence in cures, 123.

25. Bukharī 76.60–61.

26. Bukharī 76.3–4, although the Prophet discusses these forms of healing in numerous reports.

27. Bukharī 76.15, 76.1–11.

28. Bukharī 76.82–83; al-Suyūṭī, *Ṭibb al-nabi* 155–57. Jung notes that many contemporary Yemenis fumigate their homes at dawn and afternoon prayer times as protection against *shayṭān* and *jinn; An Ethnography of Fragrance,* 49.

29. Bukharī 76.79–80.

30. Qur'an 17.82.

31. Al-Suyūṭī, *Ṭibb al-Nabi* 154–57, for example.

32. Qur'an 24.61 suggests that the blind, lame, and sick are not at fault for their circumstances (i.e., sin). See Michael Dols discussion of some differing medical paradigms in the Christian and Islamic world, dealing particularly with mental illness: "Insanity in Byzantine and Islamic Medicine," *Dumbarton Oaks Papers* 38 (1984): 135–48. Also, note `Alī b. Sahl Rabban al-Ṭabarī's early medical compendium *firdaws al-ḥikma (Paradise of Wisdom)* which includes a survey of Indian medicine in the final chapters

33. Ibn Sahl's pharmacology guide exists in at least six manuscripts. The text used here probably date to c. 1040 CE; it is a revised and abridged version copies by the physicians of the `Aḍudī hospital in Baghdad. This text as well as the earlier edition have been edited, analyzed, and translated by Oliver Kahl.

34. Ibn Sahl 3.58, 7.132 (pp. 143, 164), among others. Also, see Kahl's footnote regarding Dioscorides, p. 143.

35. Ibn Sahl 1.12, 13 (p. 124); 3, 41–61 (pp. 135–45).

36. Ibn Sahl 3.58 (among others); al-Kindi, *Aqrābādhīn* 294. Also, Ibn Sahl includes recipes for stimulating sexual desire (13. 185, 187).

37. *The Syriac Book of Medicines: Syrian Anatomy, Pathology and Therapeutics in the early Middle Ages, with Sections on Astrological and Native Medicine and Recipes,* ed. and trans. Ernest A. Wallis Budge (St. Helier: Armorica Book Co., 1913), v. 2, 150.

38. *The Syriac Book of Medicines*, v. 2, 656.

39. *The Syriac Book of Medicines*, v. 2, 659, 665, 674 (among others).

40. Ibn Sahl 9, p. 123.

41. Al-Kindi, *Aqrābādhīn*, 11, p. 40.

42. Ibn Sahl 15, pp. 197–207, for example.

43. Cooking manuals also dealt with matters of contraception and sexual stimulation; see al-Warraq's recipes, chapter 125. Al-Warraq carefully warns against giving the such stimulants to women, however, because they could lose control of themselves.

44. Ibn Sahl 16.245–47, p. 212.

45. Ibn Sahl 16.248, p. 212. As Kahl points out, Ibn Sīnā also includes recipes for vaginal "narrowers."

46. Al-Kindi, *Aqrābādhīn* 188, p. 190. Al-Kindi does note that his formula will cause women to become lustful, which will more likely lead to pregnancy. This appears to be a side effect, not the goal.

47. See particularly his *Ḥāwī: Kitāb al-ḥāwī fī al-ṭibb* (Osmania Oriental Publications, Bureau, Haydarabad, 1960); discussed in Musallam.

48. There is a wide spectrum of juridical opinions about contraception and (early) abortion of course. One general guide was that any contraception that kept the semen from reaching the egg was permissible (e.g., al-Ghazālī). Abortion was permissible before God "ensouled" or "quickened" the embryo, around the fourth month (e.g., Hanafi school). This view coincides with teachings of some Hebrew Scripture and Aristotle; see John Riddle's discussion in *Contraception and Abortion from the Ancient World to the Renaissance* (Boston: Harvard University Press, 1992). Also, see B. F. Musallam's discussions in *Sex and Society in Islam: Birth Control before the Nineteenth Century* (Cambridge, UK: Cambridge University Press, 1983).

49. See Rebecca Flemming's excellent discussion in *Medicine and the Making of Roman Women: Gender, Nature, and Authority from Celsus to Galen* (Oxford: Oxford University Press, 2000), particularly pages 161–72.

50. See Riddle's discussion, 90–106. Flemming disagrees and argues instead that recipes aimed at inducing menstruation could be intended for a variety of purposes, including making conception possible. See particularly 162–63.

51. Pliny 20.51.143.

52. See Riddle, 90–91, discussing Marcellus Empiricus; and Musallam, 77–83.

53. Al-Rāzī 4, 13, 24, 19; Musallam 77.

54. Al-Rāzī 14, 21; Musallam 77.

55. Al-Kindi, *Risālat* 34–35.

56. See Martin Levey's introduction to Ibn Māsawaih, 395.

57. Ibn Māsawaih, "musk"; Levey, 399.

58. Ibn Māsawaih, "ambergris"; Levey, 400.

59. Ibn Māsawaih, "aloe"; Levey, 401.

60. Ibn Māsawaih, "camphor"; Levey, 402.

61. Ibn Māsawaih, "saffron"; Levey, 403.

62. Ibn Māsawaih characterizes several scents as appropriate for females or males and females. None of the aromatics in this selection are labeled appropriate for males only. Laudanum, for example, is appropriate for men and women's incense; Levey 408.

Part II

Sacred Scents

In late antiquity and the early Middle Ages, olfaction informed how individuals located themselves in society, usually signaling class status, morality, and appropriate gender expectations. Beyond culture and class markers, however, scent also mediated between human and Divine. Smells, both fragrant and foul, permeated the environment, transgressing the margins of private/public and transforming boundaries of secular/sacred. Religious ritual condensed time into pungent moments of symbolic unity where celebrants and deities alike received sustenance—and to deny one their scent offering was to deny the other.

Odors are particularly potent within religious ritual and ceremony because they are at once radically individual (recalling personal memories and emotions more powerfully than any other sensory stimulus) and communal (binding a group together through a shared sensory experience). Distinctive odors can indicate liminality, the transitional place betwixt and between social and theological constructs, such as sacred and profane space; salvation and damnation; or historical time and eternity. When Christians and Muslims encounter a Mass or Friday sermon, for example, smells transform space and time: it sets apart the sacred from the outside world and ordinary time collapses into a glimpse of eternity. Sweet fragrances of flowers evoke images of paradise. The effusive smell of roses wafting from a Christian corpse confirms the saint's location *between* heaven and earth: the corporal form still bound to this world while the spirit resides in Paradise. Both Christian and Muslim audiences recognize the bodies of martyrs and saints—tied to both this world and the next—after smelling glorious odors symbolic of their virtue and sanctity.

In Part II we will explore such connections between sacred and scent in the historical and cultural contexts of early Christianity and Islam. Part II consists of three chapters: the final two highlight specific Christian and Islamic sensory encounters with the Divine concerned particularly with *transformation*—the movement, or change, from one state to another. In Christian tradition, we will focus first on the particular rituals of baptism and Eucharist, primary conversion practices in the late antique and early medieval Church marking the renovation from sinner to saved. We will also examine how early Christian martyr texts employed the entire sensorium to reenact the Eucharist in a liturgical drama—as "wicked men" tortured Polycarp (d. c. 155 CE), for example, his body looked like bread and smelled of spices and frankincense.[1] Finally, we will review how Christians celebrated the lives of sweet-smelling martyrs and saints in rituals that prized their contact relics and, after death, their dead body parts; these tangible items often conferred healing and a wide array of charismatic miracles. Certainly, as we saw in Part I, death rituals included anointing bodies with spices and aromatics; yet the presence of sweet scent at death also signaled transformation and renewal—a body transformed because the soul then dwelled with the Divine.

In Islamic texts, on the other hand, God does not require such a transformation through asceticism or belief in Christ's salvific act because the body, complete with all its senses, is not the locus of sin in need of spiritual renovation. According to Muslim theology, human beings encounter Allah and enjoy His creation through their senses. The flesh and sensual pleasures, far from being associated with primeval sin or obligatory redemption, number among Allah's creation and, therefore, His wondrous gifts. Instead of focusing on the wretched sinful body transformed through Christ, Muslim authors concentrate on humanity's constant slippage from purity to impurity and back again; throughout the day, legal pollution occurs via biological functions of elimination, excretion, or emissions. After ritualized cleansing (either *wudu'*, lesser ablutions, or *ghusl*, complete lustration), the body returns to its pure state, again able to commune with God. Paradise promises an existence filled with sensual pleasure without the constant threat of pollution. Sacred scents do not signify a body's spiritual transformation as in Christianity but a body perfected in its purity. Thus, Islamic texts concerned for sensory encounters with the Divine take a very different account of the body and its constant movement between states of impurity and purity.

Before turning to Christian and Islamic sources, however, we will first examine some Greco-Roman and Jewish religious traditions that provide a context for later Christian and Islamic interpretations and cultic practice. We will focus on the general cultural milieus and the communities' religious symbolism that provide insight into early Christianity and Islam. Here, too, we find that scent often signals movement or transition—such as space (e.g., public/private) or life cycle events (e.g., marriage).

NOTE

1. *Martyrdom of Polycarp*, XV.

Chapter Four

God's Nostrils: Greco-Roman and Jewish Contexts

Multiple religious identities existed in Imperial Rome at the time of Christianity's advent, none of which fit neatly into modern notions of religiosity.[1] A plethora of Roman deities exhibited multiple qualities and personalities; Romans, for example, could worship distinctively *Venus Genetrix* (Mother Goddess), *Venus Verticordia* (the Changer of Hearts), or *Venus Victrix* (Victory Goddess). While conceptions of the divine were varied and multifaceted, ritualistic practices were more standard. Sacrifices to the deities, in particular, might have signified different things to different supplicants, yet the ceremonies themselves tended to follow a common procedure. Without proper ritual performance, Romans knew that the god might or might not respond to prayers.

Romans enacted religious rituals in a variety of venues. Seemingly private rituals took place in the home, usually ordered by the *paterfamilias*, or father of the family. These rites included sacrifice to various personal deities as well as those of the household. Particularly in Italy, Romans worshipped the *lares*, or domestic defender deities; *penates*, deities of the pantry or ancestral cult; and the family's *genius*, a divine spirit associated with family members, especially the mother (f. *juno*) and father. Most often, sacrificial offerings presented to the deities included flowers, incense, wine, or food (meal, bread, etc.); they could be held at either wall shrines or rooms devoted to cultic activity. The playwright Plautus (d. c. 184 BCE) gave voice to a disgruntled *lar* under whose niche the householder buried a pot of treasure. As the householders grew increasingly negligent of their deity, he too denied them until a pious granddaughter began to "pray to me with incense, wine . . . or makes me garlands."[2] For her sake, the *lar* once again blessed the *paterfamilias*.

More public religious expressions included sacrifice to or celebration with multiple civic deities, one's preferred deity, the emperor, or even nature itself. Fragrance and incense, as "bloodless" offerings, inundated most of these experiences. The gods and goddesses, according to traditions, particularly enjoyed sweet-smelling sacrifice because they themselves smelled sweet. Their aroma, sometimes called ambrosia, marked their presence; the divinities even fed upon such sweet fragrant offerings.[3] Venus' hair, for example, smelled of ambrosia;[4] and she used the divine unguent to revive Aeneas (a founding hero of Rome).[5] While wealthy families accessed an array of sacrificial spices and perfumes, frankincense could be found fairly cheaply because almost everyone used it in offerings. Romans even scented the gods' statues and temple walls for the divinities' pleasure.[6]

Sacrifice to the divine required an altar but not necessarily a temple, and many altars were even portable. Celebrants presented both bloodless offerings, most commonly used for thanksgiving and supplication, and bloodied (*immolation*), with a domestic animal's sacrifice. For such formal offerings, priests/priestesses approached an altar along with the *victimarii* (the one who performed the slaughter): the *tibicen* (or musician); and usually *camilli* (attendants). These assistants, usually children, carried the incense box and sacred utensils.[7]

Displays of such reverence or piety generally included wine libations and incense gifts along with the flesh. The sacrificial beast would be split open; if the internal organs presented no blemish, they would be removed and burned upon an altar, with the roasted meat divided among participants. Romans recognized that, on occasion, the gods attended the rituals; other times, the scent of sacrificed flesh and incense wafted heavenward to the gods' nostrils, ears, and mouth. Roman satirist Lucian (d. c. 200 CE) provides an important allegory in understanding this transmission. He described the divine joy of sacrificial scent in *Icaromenippus*, a dialogue depicting Menippus' journey into the heavens to meet the gods. He sees Zeus approach a line of holes with lids on them; the god then inclines his ear to the first hole, listens to the prayers of supplicants, and then responds to each. The third hole opened to sacrificial smoke, revealing the name of each worshipper; Zeus then responds appropriately. Menippus joins various gods at the dinner table where they enjoy ambrosia and nectar, and "particularly enjoy feeding on the savoury smoke of sacrifices which comes up to them and on the blood of the victims, which sacrificers pour around the altars."[8]

Roman sacrificial rituals revealed a somewhat symbiotic relationship between the worshippers and the divine. This type of relationship extended beyond what we might call the religious realm—a principle called *pietas* bound Romans to each other, their families, their countrymen, and their Rome. Roughly translated as piety, the term suggests humble veneration. *Pietas* more broadly understood applied to the relationships between parents and children; gods and worshippers; Roman patrons and their clients. Patrons, for example, expected their clients (such as poets) to worship their *genius* especially at birthdays; and to display devotion with sweet cakes, floral wreaths, incense and unguents.[9] Such reciprocal responsibilities defined social bonds, usually long-lasting and difficult to sever; and, they provided a sense of continuity to Roman existence. Many Romans, for example, inherited the rights of *pietas* within their family as well as their ancestral deities.

On one level, this relationship was pragmatic: humans, through their sacrifices, consecrated pleasing donations to their providers—both human and divine—in return for favor and preservation. Yet no clear-cut division of private/public emerges; an individual's failure to properly venerate the divine could result not only in personal injury but also public famines or military failure. When Romans—each one individually responsible—failed to assuage the deities of Rome, divine retribution could occur. Roman leaders essentially forbade their citizens to neglect the deities/patrons.

Emperor worship also allowed for relational exchange akin to divine adoration. Throughout the Roman Empire, citizens venerated their leaders after death as they joined their ancestors and famous heroes. The poet Ovid (d. c. 18 CE) suggested that Augustus Caesar, if rightly worshipped after death, could hear prayers and grant supplications.[10] Should an emperor fail to bless his supplicants, however, the state could withdraw his (and his family's) divination. Some emperors were also worshipped—as their *genius*—while living, symbolically serving as the Roman peoples' *paterfamilias*.[11] The emperor's *genius* received sacrificial offerings in public state cult as well as private worship. Household emperor worship tended to focus on the current ruler instead of past leaders.[12] Pliny the Younger, for example, asked the emperor's permission—both Nerva and then Trajan—to add their statues to a temple he planned to build.[13] Ovid wrote that he offered incense and prayers to various emperors' portraits, updated with the latest ones, every morning.[14] Such a relational exchange between human and Divine (deities and emperors alike) features prominently in later Christian and Islamic cultic practice.

The question of scent in Jewish religious ritual is a bit more complex. Ancient biblical texts certainly celebrate aromatic offerings, pleasing to God. Jewish sacred texts described temple sacrifice replete with complicated formulae for incense offerings. One incense—composed of myrrh, cinnamon, cassia, and calamus—anointed sacred space, including the tent, ark, altars, and even the officiating priests (Exodus 30:22–33). The scent secured the priests a unique authority, the ability to mediate between the people and YHWH. Another scent—used only for YHWH, a blend of fragrant spices and frankincense—distinguished God from his priesthood (Exodus 30:34–38). Anyone who employed these particular formulae for another purpose, even luxury, would face exile (Exodus 30:33, 38).

An incense altar stood in front of the Holy of Holies, the most sacred area of the Jewish tabernacle/temple, and priests burned the sacrifices twice a day (morning and evening).[15] God dwelled in the Holy of Holies, and His unique scent effectively distinguished this space from others. Here, the scent radically marked the sacred from the profane, preserving an absolutely pure space—and "soothing space"—for God to inhabit. This liminal area, between the adytum (where YHWH dwelled) and court (general priests), represents the in-between, the convergence point between humanity and Divinity.[16] As with the unique priestly scent that signaled authority and access to the sacred, here again scent intimates proximity to divine presence. The priests who entered the space not only anointed themselves with their own unique fragrance but also followed strict routines of physical purity. Any smells—sweat, menstruation (by contact), or sexual scent—would transgress sacred boundaries. YHWH's space perpetually smelled of potency, vitality, and joy instead of the sickness, decay, and sin associated with everyday life.

The Jews' God also delighted in animal sacrifice that He considered a "pleasing odor," which provides a glimpse into ritual process that delights and affects both celebrants and deity. Leviticus provides detailed instruction for the priests, requiring male offerings without blemish. Priests generally accepted these animals from among the people's livestock, including bulls, sheep, goats, or fowl. The priests "turned [the animals] into smoke" upon the altar; transformed into a fragrant offering, they wafted to God's nostrils.[17] They also provided grain offerings to God, including "choice grains," unleavened or cooked cakes. Some could be mixed with oil and frankincense; others, only with oil. This might be a concession for the poor, who could afford neither animal sacrifice nor frankincense.[18] Such practices complemented other ancient Mesopotamian and Egyptian sacrificial systems requiring material sacrifices (scent and animals) to material gods. As with Rome, the sweet smells of incense and animal sacrifice pleased the Divine, earning His pleasure, forgiveness, and gratitude. Even King David compared his prayers to sacrificial aroma, hoping that they would rise to God's presence like incense (Psalm 141:1–2).

These texts suggest that the God of Israel required a sweet-smelling space in which to dwell (the Holy of Holies) and enjoyed the aromatic offerings of burnt sacrifice, grain, and incense. This seems paradoxical as Jews eventually distinguished themselves from their polytheistic neighbors by claiming their God transcended material limits, thereby rejecting idolatry as fiendish ceremony. Deuteronomy 4, for example, cautions the Jews against falling into the error of idolatry, worshipping "man-made gods of wood and stone, which cannot see or hear or eat or smell" (Deuteronomy 4:28). Even though Biblical authorities warn against anthropomorphizing the divine, they nonetheless refer to their God's body parts. In Genesis and Exodus, God speaks existence into being; sees that His work is good; blasts the Red Sea's waters apart with air from his nostrils; and extends His hand against His enemies. Unlike those of false gods, however, YHWH's sensory organs actually worked; and, most particularly, He could smell His people's sacrifices and He delighted in sweet scents.[19]

Such texts, which inaugurate the proper rules of sacrifice and tabernacle/temple worship, established the priestly cult's central authority. The priestly texts emphasized God's pleasure in burnt sacrifice and strict purity regulations surrounding sacred space as well as the body, particularly the priests', as they could only encounter YHWH in a pure state. The high priest performed an annual sacrifice for the entire community within the Holy of Holies (described in Leviticus 16) while other priests performed daily sacrifices on outlying altars.[20] Thus many priestly texts emphasized purity of body: for example, priests followed strict regulations relating to marriage, bodily emissions, clothing, food, and contact with the dead. Their purity and attention to detail affects the rituals they perform before God. The priests' bure bodies insure, indeed even mandate, ritual efficacy through sacrifice (both bloody and bloodless).

Prophetic texts such as Jeremiah and Isaiah, on the other hand, claim their authority not only through priestly cult but also through personal inspiration and response.[21] Prophets, in contrast with priests, would require few rules for inclusion and proofs of physical purity. In critiquing the priestly cosmology, prophets also question (although do not completely discard) the efficacy of temple ritual, replete with sacrifice and purity obligations.[22] They argued that God no longer took joy in the sweet smells of sacrifice but found them repugnant either because of the unrepentant Israelite or illicit cultic practices. According to Isaiah, God had grown sick of sacrifice: "Trample my courts no more; bringing offerings is futile; incense is an abomination to me" (Isaiah 1:13). Some prophetic texts emphasize the foreign, "otherness" of incense to native Israel: "Of what use to me is frankincense that comes from Sheba, or sweet cane from a distant land? Your burnt offerings are not acceptable, nor are your sacrifices pleasing to me" (Jeremiah 6:20).[23]

The Prophet Amos proclaimed even more adamantly, as God said: "I hate, I despise your festivals, and I take no delight in your assemblies. Even though you offer me your burnt offerings and grain offerings, I will not accept them: and the offerings of well-being of your fatted animals I will not look upon" (5:21–22). Instead of burnt sacrifice, the prophets stressed that God wanted a contrite heart and pure deeds. In Isaiah (1:16–20), for example, God directed His people to

> wash yourselves [from sacrificial blood]; make yourselves clean; remove the evil of your doings from before my eyes; cease to do evil, learn to do good; seek justice, rescue the oppressed, defend the orphan, plead for the widow.

Isaiah 66 indeed places YHWH's election and sublime presence not with His priesthood but with those in distress, those who "trembled for God's word," and the "poor and broken of spirit."[24] Thus prophetic texts reimagine the hierarchical authority and access to power inherited in the priestly lineage.

Ezekiel 16 offers a sharp criticism of Israel—and her priesthood—as an adulterous wife. Prophetic texts exhibit Israel as an economic, political, and religious polity in their censures because all Israelites stood together before YHWH as a "chosen people." When unrepentant, Israel, instead of the loving and faithful wife, became a harlot who lusted after other men, or foreign cultures and their gods. As a whore, Israel profaned God's victuals and offered them to her foreign lovers (false gods) in an offensive feast:

> You also took your beautiful jewels of my gold and my silver that I had given you, and made for yourself male images, and with them played the whore; and you took your embroidered garments to cover them, and set my oil and my incense before them. Also my bread that I gave you—I fed you with choice flour and oil and honey—you set it before them as a pleasing odour; and so it was, says the Lord GOD. You took your sons and your daughters, whom you had borne to me, and these you sacrificed to them to be devoured. As if your whorings were not enough! You slaughtered my children and delivered them up as an offering. (Ezekiel 16:17–21)

Ezekiel not only criticizes the efficacy of temple sacrifice itself but questions Israelite fidelity in its cultic practice—including their most sacred rituals of food and scent offerings ordained by YHWH.

Competing views of Jewish sanctity and ritual between priestly and prophetic voices endured throughout the Second Temple Period (c. 536 BCE–70 CE) while priests continued to burn spices at the Holy of Holies' Golden Altar as well as make animal sacrifices. According to Josephus (d. 100 CE), high priests were still anointed with cinnamon-scented oil in his own time. After the Roman's destruction of the temple in 70 CE, however, Jewish authority fractured even more.

Jews in the Diaspora (post-70 CE) talked about sacred scents differently. With the destruction of the temple cult and public aromatic rituals, fragrance remained consigned mostly to private/domestic space. Even today, for example, Jews mark the Sabbath's end in the home with a ritual blessing over wine, spices, and light called the Havdalah. The prayer, which encourages Jews to engage all the senses, includes: "Blessed art thou, O Lord, our God, creator of the universe, creator of all kinds of spices." While the exact origin of the prayer remains unknown, it might speak to the role aromatic offerings once played in temple life.

Separate from the practical use of aromatics in domestic rituals, Jews symbolically correlated their virtuous lives and even ascetic acts (e.g., chastity and fasting) to incense offerings. This correlation between sweet scent and devotion was innovative yet also grounded in vital lessons from prophetic texts. Living Jews, pure and pious, constituted sweet "sacred scents" instead of the formerly required temple sacrifices. Rabbinic scholars appropriated aromatic imagery in crafting their idealized models of virtue, almost always complicated with varied gender expectations.[25]

In *Pesiqta de Rab Kahana*, a midrash that overlaps with *Leviticus Rabbah*, the rabbis poetically relate the importance of religious virtues, focusing primarily on males. The rabbis rank pious males according to the flora in Leviticus 23:40, a description of Sukkot (the Festival of Booths): "And you will take for yourselves on the first day fruit of the Hadar tree, palms of date palm tree, and boughs of leafy trees and willows of the stream." The midrash correlates "good" Jewish men with the Hadar tree's fruit: "Just as the etrog has scent and [it can be] eaten, so too Israel have men who are masters of Torah and doers of good deeds."[26] The authors then correlate the palms and boughs with men of middling virtues. The lowest ranking men, like the willow branches, remain inedible and have no scent; i.e., they know no Torah and perform no mitzvoth.

Many midrashim also draw upon the Song of Songs and expressively describe the rabbis' fragrant lives, plucked like flowers from the garden/world.[27] Rabbinic attitudes toward women contain more rebuke, however. While midrash and Talmud suggest many positive feminine role models, they are tempered by traditions that relate women with cosmological guilt and a corrupt nature that distinguish them in a fundamental way from males.

We find one such critical complaint in *Genesis Rabbah* 17.8. The text explores why women wear perfume more than males:[28]

> They asked R. Joshua . . . "And why does a woman need to perfume herself, but a man does not need to perfume himself?" He said to them, "Adam was created from earth, and earth never decomposes. But Eve was created from bone. For example, if you leave meat three days and it is not salted, it becomes putrid."

In this passage, the rabbinic author casts suspicion on women wearing perfume not because it distinguishes wanton whores. Eve is simply "other," secondary to God's primary creation, Adam. The text associates Adam/male with life; Eve/female with death/putrefaction. In doing so, rabbis remind women of their existential difference as well as their foremother's sin/guilt.

Of course other Jewish rituals concerned the relationship between humanity and the Divine that manifest on the male body. *Genesis Rabbah* 47.7 describes the first family circumcision, the deeply symbolic ritual connected with covenant, collective guilt, and redemption.

> R. Aibo said, "in the hour that Abraham circumcised the children of his house, he erected a hill of foreskins. And the sun shone down, and they became

> worm-eaten. The smell of them went up before the Holy One, blessed be He, and it was like the ingredients of the incense offering. The Holy One, blessed be He, said, "In the hour that my sons come [to do] transgressions, I will remember for them that smell, and I will be filled with mercy toward them."

Herein the foreskin becomes the incense offering, one that effected YHWH's mood and evoked mercy. God also pledges "to remember for them that smell," a powerful image that realizes the connection between memory, scent, and emotional response.[29]

Scent in the Greco-Roman and Jewish cultural contexts thus activates religious ritual that bridges the chasm between celebrants and the Divine, ruptured through sin and disobedience. In Roman and Jewish tradition, sweet fragrance both soothes and satisfies celestial inhabitants. God/s' "nostrils" inhale the aromas of incense, grain, flowers, and immolated meat, prompting benevolence. In addition to ritual time, aromatics sanctify sacred space by marking transitions between private/public and holy/profane. Scent commemorates relational exchange by distinguishing "where" the Divine resides, "who" has access, and "whom" shall be blessed; and these categories hardly remain stable. Following the "scent," and transitional moments it signals, helps us understand early Christianity and Islam's obsession with sacred scents to which we now turn.

NOTES

1. See Charles King's article describing various Roman beliefs about the gods in "The Organization of Roman Religious Beliefs," *Classical Antiquity* 22.2 (October 2003): 275–312.
2. Plautus, *Aulularia*, prologue; see Palmer Bovie's translation in *Plautus: The Comedies, V. II*, eds. David R. Slavitt and Palmer Bovie (Baltimore: Johns Hopkins University Press, 1995), 96–97.
3. See Caseau, 118–121; although it is largely outside the scope of this work, see Marcel Detienne's, *Gardens of Adonis: Spices in Greek Mythology* (Princeton, NJ: Princeton University Press, 1994).
4. Vergil, *Aeneid* I, 402; also discussed in Caseau, 219.
5. Ovid, *Metamorphosis* 14, 605–7; also discussed in Caseau, 222.
6. Saara Lilja, *The Treatment of Odours in the Poetry of Late Antiquity* (Helsinki: Societas Scientiarum Fennica, 1972), 34–35; also see Classen, Howes, Synnott, 18, 46.
7. I. C. Mantle explores the role of children in sacrificial acts in "The Roles of Children in Roman Religion," *Greece & Rome*, Second Series 49.1 (April 2002): 85–106.
8. Lucian, *Icaromenippus* (High Above the Clouds), trans. C. D. N. Costa, *Lucian: Selected Dialogues* (New York: Oxford University Press, 2005), 57–58.
9. Tibullus, *Carmina* 1.7.51–4; discussed in Bowes, 28–30.
10. Ovid *Ex ponto* 4, 9; 4, 13, 24; *Met.* 15,869 f.; discussed in Duncan Fishwick, "Prudentius and the Cult of Divus Augustus," *Historia: Zeitschrift für Geschichte* 39.4 (1990): 483–84.
11. See Ittai Gradel's *Emperor Worship and Roman Religion* (Oxford: Clarendon Press, 2002). Gradel argues that worship of the emperor's *genius* solidifies only in the later second century (i.e., not with Augustus).
12. See Gradel, 199–201. He argues that domestic worship of the emperor is largely absent from archeological remains because of their temporary nature (e.g., small statues instead of mosaics or expensive floor decorations).

13. Pliny, *Ep.* 10.8, 3.4, 4.1; discussed in Gradel, 202.
14. Ovid, *Ex ponto* 2.8, 4.9; discussed in Gradel, 203.
15. A particular fragrance was reserved for YHWH, requiring a special combination of incense items. This signals a special sacred space, marking YHWH's own dwelling. See C. Houtman, "On the Function of the Holy Incense (Exodus XXX 34–8) and the Sacred Anointing Oil (Exodus XXX 22–33)," in *Vetus Testamentum* XLII.4 (1992): 458–65.
16. See Deborah Green's discussion in her fine work, *The Aroma of Righteousness* (University Park: Pennsylvania State University Press, 2011), 75–76.
17. Menahem Haran suggests that the unique verbal use here might mean that the priests added incense to the fire to regulate the flame and enhance the smell of burnt flesh, thus rendering it a "pleasing aroma" to God. See "The Uses of Incense in the Ancient Israelite Ritual," *Vetus Testamentum* 10.2 (April 1960): 113–29.
18. See Philo's discussion of humble offerings in *On the Special Laws* 1.271.
19. John Chrysostom (d. c. 407 CE) later criticizes the Jews for this theological inconsistency and argues that "*God has no nostrils* but is a bodiless spirit. Yet what is carried up from the altar is the odor and smoke from burning bodies, and nothing is more malodorous than such a savor. But that you may learn that God attends to *the intention of the one offering the sacrifice* and then accepts or rejects it." *Against the Jews*, Homily I.VII.3 (emphasis added).
20. The incense used inside the Holy of Holies during the annual sacrifice for the community provided more than just a sacred place for YHWH to dwell. The heavy smoke shrouded the Mercy Seat, which no human (even the High Priest) should view, upon the threat of death. Deborah Green points out that many scholars have argued regarding techniques, timing, and purpose of incense smoke during the Day of Atonement rituals. She emphasizes the practical purpose of soothing God and putting Him in a "good mood." See Green, 75–76.
21. See the fine article by Ronald S. Hendel, "Prophets, Priests, and Efficacy of Ritual," in *Pomegranates and Golden Bells: Studies in Biblical, Jewish, and Near Eastern Ritual, Law and Literature in Honor of Jacob Milgrom*, eds. David P. Wright, David Noel Freedman, and Avi Hurvitz (Winona Lake, IN: J. E. Eisenbrauns, 1995), 185–98. Throughout his argument, he relies upon Weber and Mary Douglas as categories/types in deciphering priestly and prophetic authority.
22. The last chapters of Ezekiel imagine the reinstitution of the Jerusalemite cult, and Haggai and Malachi critique the Judahites who have abused the cult.
23. Also discussed in Green, 106.
24. See also Green's discussion of Isaiah 66, 104–5.
25. Deborah Green notably traces several other important "aromatic themes" that Rabbis employ including "Theological Geography" and formulating "the other" in chapter 4.
26. *Pesiq. Rab. Kah.* 27.9, as discussed in Green, 148–49.
27. See Green's discussion, 170–74.
28. See important discussions of this passage in Judith Baskin, *Midrashic Women*, 65–68; Green 136–38.
29. *Songs Rabbah* 4.6 also further correlates the hill of foreskins that transformed into an incense offering with the "mountain of myrrh and the hill of frankincense" noted in *Songs* 4.6. See Green's discussion, 188–92. There is no standard ritual use of aromatics at circumcision ceremonies that I have found. Elisabeth Hollender, in "The Ritualization of Circumcision in Medieval Judaism in Relation to Islam and Christianity: An Overview," *Religion* 42.2 (April 2012): 233–46, does discuss the exchanges between Christian and Jewish polemicists surrounding circumcision and baptism rituals. Much of the Christian rhetoric focused on the central role of Elijah's chair instead of ceremonial scent.

Chapter Five

Transforming the Body: Scent in Christianity

Scent marks important moments of transition and liminality in Christian tradition, much like the Roman and Jewish practices discussed earlier; here, we will turn our focus to a more careful review of what it signals, particularly with the ritualized body. We will first examine the transformations that scent indicates in two important Christian customs, baptism and the Eucharist, and see how they connect with demonic possession and martyrdom. We will consider the rituals' gendered ideals so profoundly based in the body (both human and Divine) and then finally look at early and medieval paradigms of the perfected Christian form.

Scent indeed becomes one of the earliest symbols in distinguishing early Christian identity against its Greco-Roman and Jewish counterparts. The New Testament, especially in Hebrews, stressed that the temple cult perpetrated by Jewish priests supplied but "a shadow of the good things to come and not the true form of these realities" (Hebrews 10:1–2). The offerings made in the temple, both incense and blood, had once served to atone for the sins of the people, but those offerings were imperfect and thus continuously repeated. God required a perfect sacrifice; and, according to early Christianity, had provided it in His son, Jesus. Jesus' blood sacrifice afforded perfect atonement, allowing human beings to be "sanctified through the offering of the body of Jesus Christ once for all" (Hebrews 10:10). To explain how God no longer took pleasure in temple cult, New Testament authors also drew upon prophetic texts that described burnt sacrifices as abominable to God. While other Jewish leaders had recognized these warnings as admonitions for virtuous conduct, Jesus' followers identified these teachings as indicators of a New Covenant, one completed in the Messiah's salvific death.

In describing Jesus' flawless sacrifice, Paul (or Pseudo-Paul) drew upon temple imagery, equating His death to a "fragrant offering and sacrifice to God" (Ephesians 5:1–2). Yet Jesus was not the only one likened to the temple offering; Paul also encourages Jesus' followers "to present your bodies as a living sacrifice, holy and acceptable to God (Romans 12:1). In this way Paul asserts himself as a new high priest, arbitrating between God's expectations of righteousness and the offerings' suitability. This is even more explicit in his letter to the Philippians, when he approves the gifts sent by Epaphroditus as "a fragrant offering, a sacrifice acceptable and pleasing to God" (4:18). Paul assures the congregation that God will thus reward their sacrifice and "satisfy every need of yours according to his riches in glory in Christ Jesus" (4:19).

Paul's most poignant use of scent is found in 2 Corinthians. Here, he employs fragrance as a symbol of the spiritual conversion so important throughout his theology. In Romans, Paul introduces the notion of transformation and directs Jesus' followers to "not be conformed to this world, but transformed" into acceptable sacrifices to God (12:2). In 2 Corinthians, Paul compares such rejuvenated creatures, triumphant through Christ, with sacrificial odors. He writes:

> But thanks be to God, who in Christ always leads us in triumphal procession, and through us spreads in every place the fragrance that comes from knowing him. For we are the aroma of Christ to God among those who are being saved and among those who are perishing; to the one a fragrance from death to death, to the other a fragrance from life to life. Who is sufficient for these things? For we are not, like so many, peddlers of God's word; but as men of sincerity, as commissioned by God, in the sight of God we speak in Christ. (2:14–17)

Paul's correlation between triumphant Christians certainly alludes to temple sacrifice but it also references Roman Imperial processions; as soldiers returned home in triumph, fellow citizens almost always welcomed them with incense.[1] Both allusions praise victors who had struggled and won; in the first, the battle is spiritual and in the other, physical. Both also present a sense of sacred presence among the mundane—the body of the transformed Christian, aromatic to God, and the body of the soldier, vital and unbroken.[2] In Christian tradition, fragrance came to signify not only a suitable offering made to the Divine but also the Divine presence among His worshippers.

Two rituals in particular marked the spiritual transmutation of early Christians, Jew and Gentile alike: baptism and the Eucharist. This religious renovation brought about a change particularly in the body's senses. As explained first by Origen, God created the "inner" man in His image: meaning, the "outer" man consisted of matter, but the "inner" man was "immaterial and superior to all corporeal existence."[3] Origen instructs Christians to obey Paul's command in Romans 12:1—"be transformed"—and, for him, that required the recognition and purification of the "inner" man and its senses. Indeed, he postulated that each of the five senses had an alternate and more reliable counterpart. When properly developed they transmute the physical sensorium into superhuman sin detectors. The outer man had a nose, for example, but the "inner" man

> with different nostrils perceives the good smell of righteousness and the bad smell of sins. . . . We perceive with our nostrils good and bad smells in the world of sense, so also for the inward man there is a perception of the good smell of righteousness such as the apostle had, and an evil smell of sins, which is possessed by the person whose divine senses are in good health.[4]

Origen continues to describe evil as a foul and rotten stench, one easily exposed by the transformed, Christian "spiritual" nose.

Origen's rhetoric introducing the outer/material and the inner/spiritual body reveals a tension that becomes more pronounced in other texts. In late antiquity and the early Middle Ages, Christian theologians viewed the body with profound ambivalence and sometimes disgust. The body was at once the temple of God and also the source of temptation; Galatians 5:17 declared that "what the flesh desires is opposed to the Spirit, and what the Spirit desires is opposed to the flesh." Yet humanity encountered God through the body's senses, seeing and even smelling the fragrance of God. As S. Augustine notes:

> You called me; you cried aloud to me; you broke my barrier of deafness. You shone upon me; your radiance enveloped me; you put my blindness to flight. You shed your fragrance about me; I drew breath and now I gasp for your sweet odour. I tasted you, and I hunger and thirst for you. You touched me, and I am inflamed with love of your peace.[5]

Much like Origen's "inner" sensorium, this encounter between the supplicant and the fragrant God requires a salvific experience—initiated by God—that successfully transforms the body, transmuting the sinful to the spiritual, including the sensorium.

In ascetic discourse sensory inversion oftentimes accompanies this spiritual transformation, making the foul fragrant and the fragrant foul. The fifth-century Syriac saint Symeon Stylite's harsh austerity left him with a worm-infested waist decayed from a tightly wound rope, yet his body wafted the "sweet smells of asceticism." God transformed the body, reduced in various states of decomposition, into a fragrant sacrifice, even though the viewer might only outwardly perceive putrefaction and decay.[6] The ascetics, in disturbing acts of heroism, struggled to destroy their "outer" men so only the spiritual might exist.

When a group of Christian Syrian villagers converted to Islam in the eighth century, their neighbors identified them by their odor as well:

> But they [the converts] grew different from the faithful people in both person and name; in person, because their once happy appearance became repugnant, in such a way that they were recognized by the intelligent ones through their persons, odor and the look of their eyes . . . instead of a sweet odor of the holy myron a stinking and fetid smell emanated from them.[7]

Hagiographers related transformed Christian and ascetic bodies with sweet-smelling aromas and the apostates, formerly fragrant believers, with stench. This signaled the Christians' status as heirs to eternal life in Paradise, and the unbelievers as doomed to death and decay.

Church Fathers were eager to reveal this new sensorium available to their novitiates beginning with their catechist training and initiatory rituals. Many liturgical manuals and scriptural commentaries thus focus on the full sensory experiences related to two of the most important rites: baptism and communion.

BAPTISM AND DEMONS

Early Church Fathers carefully described the new sensorium available to the renovated Christians because it would require them to readjust their entire beings.[8] Priests, for example, generally kept the Eucharist hidden from novitiates who would be dismissed before the central ritual activity. Beginning with their baptism they would participate in sacraments that required them to see with their "eyes of the heart" (*oculos cordis*). Baptismal waters were not just water; they were made "sweet" after a ritual blessing. Fragranced oil anointing their bodies would provide protective armor. The Eucharistic host would taste like bread, even though it would be flesh; wine would look like wine, even though it would be blood. Employing the sensory organs—both physical and spiritual—helped Christians fathom the mystery of Christ, the incarnate Divine. The Christian's body became a vehicle for approaching God-in-Body: hear Him, touch Him, taste Him, smell Him.

Before participating in the Eucharist, however, Christian initiates had to be baptized; symbolically, they experienced death and spiritual rebirth into the mystical body of Christ.[9] Priests enjoined new Christians to receive baptism not only to imitate Jesus, as he had received baptism from John (Matthew 3:13–17), but also to mark their own spiritual transformation from sin to righteousness, death to rebirth. Several Church Fathers even imagined the initiates as children, or "little ones" (*parvuli*), emerging from the baptismal font, regenerated from their sinful form in the likeness of Christ.[10] Zeno of

Verona (d. 371) likened the pool to a mother's womb and the candidates to infants, reformed as siblings regardless of age or race.[11]

The symbolic correlation between baptism and rebirth is reified even more so in the initiates' nudity. Just as we are born naked from the mother's womb in physical birth, Church Fathers interpreted baptism as a "second birth" wherein naked initiates emerged from the baptismal font/womb remade.[12] The *Apostolic Traditions* instructs candidates to "strip naked" and then follow the Bishop or Presbyter for anointing and then water baptism.[13]

Early Christians performed baptism, preferably, in running water, but if unavailable, any water would do.[14] While many variances exist among baptismal rites in differing locales, a general order does tend to appear.[15] The customary ritual included prebaptismal questioning and exorcism; anointing with oil (probably unscented olive oil); stepping naked into the blessed waters; and a postbaptismal anointing (usually scented). Application of the (unscented) oil served several theological purposes including "putting the Devil to flight" (as a tool of exorcism); "sealing up the breast" against sin; and (scented) importing the "fragrant gifts of the Holy Spirit."

The oil used in the final ritual anointing probably contained aromatics. Ambrose explains that the baptismal initiates constituted a "sweet odor to God." This may indicate more than poetic allusion to Christian virtues and instead describe the aroma of the ritual itself. Indeed, church manuals permitted the addition of spices to the holy chrism even though they offered no standard formula.[16] Aromatic recipes today still include spices that resemble biblical formulae with cinnamon, cassia, and balsam; some even add ginger, cloves, and cardamom.[17] Zeno of Verona compared the scent of the newly baptized to freshly baked goods, immanently more sublime than the stench and filth into which they were born.

> Now our mother (the Church) adopts you so that she may give birth to you, but not in the manner in which your mothers bore you when they brought you into the worlds, themselves groaning with birth pains and you wailing, filthy, done up in filthy swaddling clothes and surrendered to this world, but with joy and gladness and freed from all your sins, and she feeds you not in a stinking cradle but with delight from the sweet smelling rails of the holy altar.[18]

After their baptism and anointing, Christians stood reborn from sin and stench into righteous fragrance.

The pre- and postbaptismal anointing not only signified the newly regenerated Christian but also the Holy Spirit's added layer of protection (or "sealing") against sin and temptation. Liturgical formulae required the initiate to renounce Satan and his "pomps," most likely referring to Roman spectacles such as gladiatorial games, circuses, and amphitheaters.[19] Augustine taught that Christians who continued to attend such displays after baptism "offered incense to demons within their hearts."[20] Indeed, no charioteers or gladiators could bathe in baptismal waters until they renounced their occupations.[21] Some rites required the anointing of the entire body (also present in gladiatorial combat, ironically) in provision against Satan's wiles; others singled out the breast, implying the heart, which dwells upon the pleasures of sin. Still other rites mentioned the head and sensory organs such as the ears and nostrils, orifices that allow for temptation to enter the body.[22] Ambrose suggests that priests anoint the ears in recognition of Christ's miraculously healing the deaf-mute (Mark 7:31–37); and the nostrils because the initiates now provided a "good odor" to God.

Sealing sensory organs also relates to exorcism, an important aspect of baptism. Baptismal ritual realized the salvific transformation with the waters' purification and healing. In preparation priests exorcised novitiates daily and then provided a final examination before the immersion; if found pure, according to the *Apostolic Traditions*, the priests then "breathed on them" and sealed them on the forehead, nose, heart, and ears.[23] The breath, or exsufflation, might refer to Jesus' own breath expressed upon the disciples, imparting the power to forgive sins (John 20:22–23). It might also relate more directly to the early Christian practice of hissing, spitting, or blowing on pagan altars or "false idols," thus resembling an act of exorcism.[24] In *Against Celsus*, Origen relates Celsus' comparison of Christians to magicians who, "for a few obols make their sacred lore in the middle of the marketplace and drive demons out of people and blow away disease."[25] Regardless of interpretation, when Christians received baptism, they could expect all of its therapeutic benefits including physical and spiritual cleansing as well as sealing from future invasion.

The correlation of Christian baptism and healing (scents) should not be underestimated here.[26] As discussed in Part I, late antique and early medieval etiologies included miraculous and celestial sources alongside more temporal reasoning located in Galenic models (among others). Jesus demonstrated his own charismatic authority in multiple healings from ailments such as blindness, leprosy, paralysis, and demonic possession; and, He passed that authority—and healing power—on to the apostolic hierarchy located within the Church. According to Luke 9:1: "Then Jesus called the twelve together and gave them power and authority over all demons and to cure diseases, and he sent them out to proclaim the kingdom of God and to heal." Indeed, early Christians associated healing (of body and soul) with demonic expulsion in general.

Holy men and women served the important function of exorcists—they rid Christian individuals and communities of demons, usually identified through their foul smell. Pope Gregory the Great alludes to the general association of evil with stench when he explains that holy men—with a "nose like the tower of Lebanon"—can sense temptation by scent long before sight:[27]

> For the thing that we do not yet perceive with our eyes, we typically anticipate by its smell. And through the nose, we distinguish between a sweet smell and a stench. What, then, could be signified by the "nose of the Church" if not the saints' prescient discernment? It is also said to be like the tower of Lebanon because the discerning foresight of the saints is so set on high that they are able to see the attacks of temptations even before they come and to stand mightily against them when they do arrive.[28]

Hagiographers and theologians such as Pope Gregory exploited a powerful cultural fear that correlated stench with moral depravity and physical/spiritual threat. Demons, as well as personal immorality, could be identified through stench.

Early Christian texts often compared the most putrid scents with heretics as well as demons because both threatened the believer's pure soul. Augustine refers to his student days, those of disbelief and vice, as those of "rolling in [Babylon's] dung as if it were spices and precious ointments."[29] Theodoret of Cyrrus (northern Syria; d. 457) consigned the great Christian heretic Arius to death in a latrine, while evacuating "the refuse from his gluttony."[30] Theodoret explains that he died a "noisome death" because of his "noisome blasphemy."[31] Heretical belief occasioned the same fear and disgust as demons.

Desert fathers and mothers inhabiting the Egyptian and Palestinian deserts provide perhaps the most significant evidence for demonic stench.[32] Demons and monastics coexisted in the barren landscape with holy men and women testing their virtue against their enemies' onslaught. Putrid smells signaled the arrival of demonic beings or exposed the "true nature" of their fiendish plots. S. Anthony once boarded a ship to pray with its monks, and he noticed a foul scent. The passengers explained it as the odor of fish and dried meat, but Anthony perceived the "truth." He beckoned a demon-possessed young man and exorcised him. As Athanasius explains, "Then everyone recognized that the stench was from the demon."[33] Another demon terrorized and humiliated a nobleman so much that he "devoured his bodily excrement."[34] S. Hilarion, a Palestinian hermit inspired to the ascetic life by Anthony, also had the gift of demonic discernment according to scent. According to his hagiographer, S. Jerome, a greedy landowner once sent Hilarion his crop's firstfruits of chickpeas (which, indeed, are known to cause flatulence). The hermit perceived such a rank smell that he recoiled from its presence and sent the peas to feed the cattle, which then began to bellow and flee. Hilarion was "enabled by grace to tell from the odour of bodies and garments, and the things which any one had touched, by what demon or with what vice the individual was distressed."[35] The presence of evil provoked disgust in possessed and audience alike.

This association continues in later Latin traditions, even in relatively course activities that can be delightful for the modern reader (as well as the ancient?).[36] Gregory of Tours, for example, reports how one man procured a vial of oil from S. Martin's tomb and applied it to healing the sick and demoniacs in his hometown. The saintly relic caused one man's "more hideous demon" to be expelled through a "blast of air through his bowels."[37] After farting out his demon, the supplicant became whole. The flatulent stench identified the demon, yet Augustine explains that flatulence itself signifies humanity's sinful nature. Because of the fall, humanity has lost ultimate control over the body, thus experiencing genital arousal, farting, and sweating not according to our "will." Only rarely might one maintain dominion over such physicality; this gives us a glimpse of the prelapsarian body yet to be enslaved to its involuntary impulses.

> Some have such command of their bowels, that they can break wind continuously at pleasure, so as to produce the effect of singing. I myself have known a man who was accustomed to sweat whenever he wished. It is well known that some weep when they please, and shed a flood of tears.[38]

Augustine carefully explains that sex would have occurred in paradise, before the fall, according to the will, without "lust" or involuntary genital stimulation. He does not consider, however, the presence—or quality—of paradisiacal flatulence.

Demons associated with such stench often carefully discerned scent themselves: according to Justin Martyr, they taught pagans how to offer incense to them (i.e., they themselves preferred sweet smells).[39] In some Jewish traditions, putrid odors and certain plants even repulsed demons.[40] In the book of Tobit, Tobias cures his wife, a certain Sarah, possessed by the demon Asmodeus. Upon entering their bridal chamber, Tobias (having been instructed to do so) takes a fish's liver and heart and burns them upon embers of incense. The demon smelled the fumes and fled to upper Egypt where an angel captured him (Tobit 3.8; 6.13, 17–18; 8.3). The demon, responsive to its malodorous surroundings, responds to Tobias' ritual act.

This particular method of demonic expulsion must have been well known as Gregory of Tours later explains how he saw in a vision his father's cure for high fever, swollen feet, and severe nerve pain. A visionary character instructed Gregory to do as described in Tobit; so, Gregory's mother sent a servant to catch a fish and then they burned the fish's liver and heart. The father immediately recovered, likening the miraculous healing to demonic expulsion, which makes sense considering ancient etiologies. In a tradition recorded in Josephus' *Antiquities* (c. 94 CE) a Jewish man, Eleazar, exorcizes a demon by holding his ring filled with aromatic roots under the demoniac's nose. Here, too, a demon falls prey to scent managed by a righteous man. In these Jewish archetypes, demons respond to fragrance in certain recipes or spells.

According to other Christian traditions, in contrast, exorcism depended more on personal charisma than amulets, charms, or spells. Jesus exorcised demons throughout the synoptic gospels, sometimes by formulae or spoken command (Mark 1:25; 5:8; 9:25). On one occasion, He transferred demons from a person into an object, or a herd of pigs (Mark 5:12–14). Jesus made it clear that he relied upon "the Spirit of God" for such power (Matthew 12:28; Luke 11:20). In the *Apostolic Tradition*, bishops exorcised Christians preparing for baptism by laying hands on them. The bishop's touch transmitted Jesus' authority over evil, perhaps also with spoken words.[41]

Indeed, the authority to expel demons widely identified the "friends of God," both living and dead; and, exorcisms occurred in a variety of ways. Often the saint's mere presence sent demons howling in pain. Once, Genovefa (b. 429), patroness of Paris, healed a throng of people with "prayers and the sign of the cross," while the demons complained that she burned them with "celestial fire."[42] Frequently, in an act reminiscent of the baptismal anointing, saints covered the possessed with oil, ritually cleansing them of demonic impurity. Genovefa performed this miraculous act many times as well; according to one narrative, she healed three wives with troublesome spirits with holy oil after hearing their husbands' complaints.[43] The miracle relates both to the oil and her personal charisma as she is the one who smooths the oil onto the afflicted. It seems the actual healing oil required a blessing by a male priest or bishop, however. On another occasion, Genovefa found her oil bottle empty and the bishop far from home. After a troubled prayer—she needed oil to cure a person possessed—she arose to find the container miraculously filled.[44]

Gregory of Tours also relates S. Martin's many miracles involving charismation with oil from his tomb. Not only did some shrines provide oil that had contacted the saints' relics directly, but pilgrims also brought their own supply and then touched it to the tomb themselves. After one pilgrimage to S. Martin's tomb, a supplicant returned home with a vial of oil. Later he was "struck by a piercing pain that he could scarcely draw a breath"; so he turned and prayed in the direction of Martin's shrine and rubbed his body with the oil. He recovered immediately. His neighbors, including the sick and demoniacs, also found relief from S. Martin's oil (although Gregory does not describe any recompense gained by the pilgrim).[45]

The prolific use of oil associated with demonic exorcism and healing is not surprising given what we know about ancient medical techniques. Through the anointing, the saint conferred not only their personal salvific charisma but also, in most cases, a healing, sweet scent. In one miracle, Radegund (still living) appeared to a *monacha* in a dream, instructed her to descend into a bath with no water, and then the saint poured oil directly onto her head and clothed her with a new garment. When the woman awoke, she was cured and still "smelled of oil in witness of the miracle."[46] According to another narrative, Radegund wore wormwood around her neck "for refreshment." A fellow nun suffering from an eye infection procured a piece that the saint had worn and "the eye was suddenly clear and bright again."[47] Even though no oil is present in this miracle, the saint's charismatic contact with the fragrant plant grants relief.

While saints and priests effectively wield the "power" of scent in their charismatic gifts, the baptismal ritual associated all Christians with sweet smells. According to Church Fathers, the new Christian dies to the old self and begins life again, having joined in the death and resurrection of Christ in a regenerated immortality. Sweet scents signal this liminal phase—between death and new life: anointed with baptismal oils, healed of death, and exorcised of demons. The new Christian then rises from the baptismal womb to begin life anew with the fragrant Bride/Church.

EUCHARIST AND AUTHORITY

After baptism, new Christians gained entrance into the mystical Bride of Christ, the Church. Most significantly, they shared in the Eucharist; this ritual, too, commemorated the transformation available through the Christian narrative, the death and resurrection of Christ. While Christians gathered to remember Jesus' salvific act, they also sought to renew their own bodies as a sacrifice pleasing to God. As time passed, and Eucharistic theology became more complex, God—through the attendant priests—even replicated the sublime transformation of death/resurrection in the element of the bread/flesh at the altar. By partaking of Christ's body as food, Christians joined themselves with their fellow Christians, a new and vibrant *communitas* available to converts.[48]

Evidence for the Eucharist reveals some early tensions between domestic/personal performance and communal experience. As indicated in the writings of Paul, especially Galatians and 1 Corinthians, Christians (Jews and Gentile alike) would gather for communal meals and share in the Eucharist (or, Lord's Supper). It is probable that these gatherings consisted of a banquet, or shared meal, and then ritual worship.[49] It is likely that incense and aromatics served a mundane function herein; either to create a pleasant atmosphere, refresh guests, or flavor the food.[50] There does not seem to be a ritualistic function for incense or aromatics while early Christians still met in small house meetings.[51]

In some instances, congregants saved some of their bread and took pieces back home for consumption during the week.[52] In these rituals, it is likely that the *paterfamilias* distributed the bread. It is also plausible that the bread conserved for such purpose might have been stored on a home altar or special niche distinguished with incense, as were other Roman home altars. These practices would have faced increasing opposition as a more central authority arose in the person of the priest and bishop. The ritualized, public performance of the Eucharist only by clerical hands did not pose a problem until after Christianity became a legitimate religious cult of the Empire.

Before Constantine's fourth-century Edict of Milan and the legalization of Christianity, theologians discouraged the use of incense in Christian public ritual. They often associated fragrance with polytheistic rituals, particularly emperor worship, and condemned its use as "polytheistic." Indeed, one of the tests put to Christians to confirm their identity required an incense offering to the emperor's *genius*. Those who failed to make the offering could face exile, corporal punishment, or martyrdom. Church leaders even impugned Christians who made private incense offerings to God, labeling them blunderers that might lead other Christians to falter. Lactantius criticized any aromatic offering, whomever the intended recipient:

> lest anyone should think that victims, or odours, or precious gifts, are desired by God, who, if He is not subject to hunger, and thirst, and cold, and desire of all earthly things, does not therefore make use of all these things which are presented in temples and to gods of earth; but as corporeal offerings are necessary for corporeal beings, so manifestly an incorporeal sacrifice is necessary for an incorporeal being. But God has no need of those things which He has given to man for his use, since all the earth is under His power. . . . What then does God require from man but worship of the mind, which is pure and holy?[53]

For Lactantius, the incorporeal Christian God had no use for material offerings such as spice, perfume, or glorious temples; the Romans and Jews had made those mistakes.

Despite Christians' own conversations regarding use of fragrance during worship, aroma swept through an equally central display of Christian identity, martyrdom. Those who endured official questioning and torture, refused to make public offerings to the emperors' *genius*, and even presented themselves voluntarily, all established a new form of public spectacle substituting "pagan" bloodshed for the pious (as discussed above in Part I). However, Christian martyrs also delivered a new form of public liturgy.[54] They transformed a public ritual of criminality into a ritual performance of triumphant piety, countering the Roman's social control by reinterpreting the drama themselves.[55]

The notion of personal sacrifice—in this case, Christian death—would have been meaningful at multiple levels. Greco-Romans, for example, venerated ancient heroes who died for noble reasons; perhaps the most famous is Socrates who accepted death instead of fleeing Athens. Roman soldiers also sacrificed their lives for others' benefit or at divine command.[56] Certain acts of suicide could also be venerated as pious sacrifice; Seneca, for example praised the deaths of Cato and Scipio as victories of the human will against failure and Fortune.[57]

Judaism, too, exalted noble death most notably in the books of the Maccabees. The books explain the Jewish persecution under the Greek Seleucid Antiochus IV and the Hasmonean family's resistance. The Jewish revolution, which secured Judea's independence for about eighty years, was won only with the noble deaths of several heroes. One elder Razi, for example, stabbed himself with his own sword and plunged into a crowd, offering himself as a sacrifice for his people (2 Maccabees 14). Another important story notes the sacrifice of a mother of seven sons, all of whom were arrested and forced to eat pork (forbidden under Jewish law). After refusing her tormenters, the mother watched all seven children tortured and killed as they consistently attested to their God and the hope of resurrection. The author then described the mother as "exceedingly admirable" as she encouraged her sons to martyrdom by "arous[ing] her female way of reasoning with male courage" (2 Maccabees 7.1–42). Like later Christian martyrs, the heroes of Maccabees also hoped for resurrection by God.[58]

For early Christians, however, martyrdom resonated with one of the most central religious rites of their community, the Eucharist.[59] The acts of the martyrs, carefully described and relayed among the *ecclesia*, dramatized the sacrificial corporal offering required for salvation in a vivid display, perhaps even more compelling than the Eucharistic rite itself. While the Eucharist reenacted Christ's physical suffering by substituting bread and wine, martyrs imitated the salvific act through their torments and death. The imitation of their savior provided early Christians an intimate union with the Divine inaccessible through many other sacrificial rites. Most Roman sacrifices, for example, focused on the reciprocal responsibilities between citizens and their deities not an ecstatic union with the Divine.

Christian martyrs routinely transformed into Christ crucified through their sufferings and then resurrection into Paradise. Their mystical union with Christ at their time of death, symbolized in their ecstatic release from pain, modeled the mystical union available to all Christians as they tasted Christ's flesh in the Eucharist. Christian communities repeated these heroic acts in the most sensual of terms, correlating the martyr's body with Christ and the hearer's body with the martyr. This particular rhetorical tool, which provided *enargeia*, or vivid description, aimed to engage the audience with the narrative in emotional terms, causing them to feel rather than just picture events in their minds. Used properly, words and language thus compelled certain virtues and truths upon an audience.[60] With their scrupulous accounts of death, suffering, and reward, Christian leaders thus instructed Christians how to imitate the martyrs' virtues.

Ignatius of Antioch, killed sometime between 98 and 117 CE, wrote several letters while on his way to die in Rome. Especially in his *Letter to the*

Romans, he relied heavily upon Eucharistic language to describe his coming ordeal. He prayed that he would be made "a libation poured out to God, while there is still an altar ready for me"; and that his body be "ground fine by the lions' teeth to be made purest bread for Christ."[61] Ignatius also hoped that Christians would gather round his slain body, singing in harmony, a poetic allusion to the Church's unity commemorated in the Eucharistic rites.

The Martyrdom of Polycarp (d. c. 155 CE), bishop of Smyrna, also presents his sacrifice in Eucharistic terms; his body becomes the "bread" of Christ. The local proconsul called Polycarp before him but he refused to recant and sacrifice to Rome's gods. The proconsul finally condemned the old man to death by fire; in this scene the Eucharist reference is at its greatest. Accord to eyewitnesses, the walls of fire formed around him like a "loaf baking in the oven" that wafted a "delicious fragrance, like the odour of incense or other precious gums."[62] This passage alludes, of course, first to the temple cult practice of incense and burnt offerings; and, at the same time conflated Polycarp's body with Christ's sweet flesh, offered for humanity.

Cyprian of Alexandria even suggests that memorials to the martyrs should be included in the Eucharistic rites themselves. He directs presbyters and deacons to note each martyr's *dies natalis* (their date of martyrdom, also their birth into Paradise).[63] S. Augustine later notes that a church was built on the site of Cyprian's death, where his disciples gathered to celebrate the Eucharist and commemorate his death.[64] While Christians could not (or would not) offer fragrant incense at public gatherings in Eucharistic celebration, their descriptions of the martyrs' sufferings, employing such sensual imagery, called to mind the smells of savory smoke and incense with each retelling.

The concentration of sensory cues in the dramatic recitation of the martyrs' sufferings calls particular attention to moments of transition and transformation. The martyr's body, often sweet-smelling even as it was transmuted under the tools of torture, signified the Eucharistic mystery, the bread of sacrifice transformed into the body of Christ. The martyr was transmuted in other ways, as well, specifically through gender imagery. Males manfully, although passively, endured pain; Ignatius of Antioch even felt "birth pangs" as he was reborn into Paradise.[65] Females actively displayed their virile strength and vigor. Hagiographers recast Blandina, a female martyr in Gaul, as a victorious athlete who endured against her tormenters' weakness after a day of tortures and combat. All of that in spite of the fact that she was an "insignificant, weak, and despised woman."[66] Recitation and commemoration of the martyrs' lives and gruesome deaths, so vibrant with sensual imagery and gender inversion, emboldened Christians to embrace the virtues now required as members of the *ecclesia*.

Just as Church leaders required new Christians to learn certain teachings before baptism, and baptism before the Eucharist, they also trained them as potential martyrs.[67] This does not mean that Church leaders expected that all Christians would be put to the test and literally face death in the arena. If individuals were questioned, however, martyr stories certainly directed Christians in how to respond, what gestures and facial impressions to employ, and how to embrace suffering.[68] The martyrs themselves, while imprisoned, even offered the Christian community access to spiritual charismatic power. While held for trial, Roman prisoners depended upon their family or friends for both physical and emotional sustenance.[69] The Church itself thus accepted the responsibility for feeding and caring for their persecuted members. Tertullian, for example, refers to "Lady Mother Church" feeding the prisoners "from her fruitful breasts, and each brother from his own means, provid[ing] for your needs in prison."[70] Tertullian even warns that too much care for the prisoners' bodies leads to a luxurious abundance that dulls the spiritual senses.[71]

Tertullian's concern that excessive food, drink, and personal attention might occur indicates that Christians considered potential martyrs as loci of spiritual power. Tertullian comments that Christians sought them out for "peace"; and martyrdom narratives display them as new authority figures.[72] In the *Martyrdom of Perpetua and Felicitas*, for example, Perpetua's intercessory prayers free her brother from a purgatory-like state of temporary suffering; on her way to the arena, a bishop and presbyter prostrate themselves before the martyrs for spiritual blessing.[73] Caring for the martyrs' bodies—alive yet destined for mutilation or destruction—provided proximity to holiness that strengthened the Christian community.

Perhaps more importantly, however, the subversive rhetoric of martyrdom promised spiritual empowerment and victory instead of pity and humiliation. Christians, as did the martyrs, should exhibit self-control and freedom from passions, virtues also extolled by Roman philosophers. The men and women who died in the arenas denied Imperial authorities their ultimate power by choosing death with dignity and honor; their trials transformed victims into heroes and stripped persecutors and proconsuls of reason. Christians played the virtuous leads in public dramas designating Roman officials and pagan spectators as bestial extras.

As Christians watched their martyrs die in the arenas, they also knew they witnessed a cosmic struggle that each believer must face, the battle featuring Christ, Satan, and the angels. With these as their models, living Christians should fight with no less determination. Many Church leaders that later became martyrs themselves wrote letters for their disciples, describing how to prepare for heroic death. Such careful instruction can be found in the writings of Ignatius, Polycarp, and Pionius. Other manuals, more cohesive in their advice for sufferers, include Tertullian's *To the Martyrs*, Origen's *Exhortation to Martyrdom*, and Lactantius' *Divine Institutes* (written during Diocletian's Great Persecution, 303–13).

These "manuals for martyrs" were so important because apostasy posed one of the greatest threats to the early community. Even though general and consistent persecution rarely occurred, Church leaders worried that only mature Christians, those ready for "solid food rather than milk," could withstand the trials of persecution.[74] If Christians came under question, or were forced to offer sacrifice to Rome's gods, they had to be rightly prepared for the ordeal. If they failed their test, denying their Christianity and offering to Rome's gods, they might lose their own soul as well as dishearten the entire Church. Not only this, but many Church Fathers understood the fragrant sacrifices and burnt offerings by pagans as food, nourishment to demons.[75] Christians must be prepared adequately for their tests for the Church's sake—its enemy must not be empowered. The lives of the martyrs who successfully resisted Satan's pleas to apostatize thus became liturgical icons, processed through the church's memories and imagination, providing models for imitation as well as veneration.

The martyrs' popularity and their symbolic connections with Eucharistic offerings reveal a growing tension between domestic and public rituals central to early Christian identity. With the fourth-century community legalized and flourishing, rituals became increasingly standardized (although always varied). Prayer, for example, represented an early Christian ritual increasingly defined through a flurry of prayer manuals and guides as early as the third century. Church Fathers such as Tertullian offered prayer advice including proper times, styles, and formulae.[76] This attests not only to the importance of prayer itself but the impetus to systematize it. Early Christians had to fully articulate the social and theological distinctions between public and private rituals, an important feature in Roman practice and Jewish temple cult.

While many rituals could occur in the home, these competed with the rites performed under a singular collective leader or authority. The rise of a central specialist—who alone could present the Eucharist, for example—caused considerable angst among theologians. One of the earliest struggles concerned the confessors' authority—those who maintained their Christianity under threat of torture and survived. By the time of Gregory of Tours (d. 594), popular confessors, no longer living in an era of persecution, often achieved their titles through ascetic self-discipline.[77] Indeed, according to the Gothic Missal (c. 700), a separate mass existed to commemorate the confessor (or groups thereof), independent of martyrs and local saints.[78] Early confessors arrogated to themselves the rights of priests, deacons, and presbyters without ordination or the laying-on of hands; their authority resided in their suffering and imitation of Christ.[79] A growing number of Church authorities resisted this explanation, however, claiming that the confessor denied Christ's own ordering of the Church.[80]

Simultaneously, the divide between public/private cult came to differentiate between orthodoxy/heresy. The emerging authority of (male) priestly specialists increasingly signified the "orthodox." Their central authority was hardly without controversy; not only did debate rage over the confessor's role but many Christians also venerated the rigors of asceticism and isolation (eremitic charismatics or "holy hermits"). Priests and bishops, in effect, had to tame the holy men and women that provided an alternative to episcopal structure. And this they did.

Athanasius, Archbishop of Alexandria (d. 373), for example, constructed the famous hagiography of S. Anthony, the charismatic recluse who fought the devil in the desert. Although replete with miracles and exorcisms, the text explains that Anthony displayed an appropriate humility for local episcopal representatives. Athanasius wrote: "Though the sort of man he was, he honored the rule of the Church with extreme care, and he wanted every cleric to be held in higher regard than himself."[81] Bishop Gregory of Tours devoted four chapters of his epic *History of the Franks* to S. Martin of Tours, a charismatic holy man who not only submitted to the Church but "reluctantly" joined its ranks.

As Christianity became a state-sponsored cult in the fourth century, Church leaders finally included incense in public space. Episcopal authority, responsible for crafting orthodox theology and ritual acts, enthusiastically integrated scent, multivalent in effect and meaning, as it solidified its power and status. Scent, if one could afford it, signaled orthodoxy and authority; and the Church could afford it. According to the *Book of Pontiffs*, Emperor Constantine designated 150 lbs. of spices to adorn various Christian altars under Pope Silvester (d. 335). Many basilica also received annual revenue in the form of spices (such as cassia and saffron), nard-oil, and balsam.[82] Further depicting the episcopal access to authority, Pope Silvester mandated that only Bishops could consecrate oil for chrismation; and, Pope Boniface (d. 422) later restricted women (even nuns) from touching either the consecrated pall (ecclesiastical vestment) or incense.[83]

Church officials standardized incense in liturgical practice also as a way to compete with its various Roman counterparts. Admittedly, the scented mass resembled Jewish temple sacrifice: just as the Jewish priest yielded the burnt offering upon the altar "by fire of pleasing odor to the Lord," the Christian priest recreated Christ's sacrifice as he transformed the Eucharistic bread into the mystical body.[84] The ritual use of incense mimicked the sacrificial victim's "pleasing odor" rising to the heavens outlined in Jewish ritual. Scent identified the moment of sublime transformation reminiscent of Jewish animal sacrifice—and the martyrs' self-sacrifice—providing again a sweet odor to God's nostrils. Early Church Father John Chrysostom (d. c. 407 CE) once criticized the Jews in a scathing homily, comparing their practices to demonic worship, and proclaiming that "God has no nostrils!" and is, indeed, a bodiless spirit.[85] Meanwhile, the Church still crafted rituals that appealed to both human and Divine sensorium and articulated a theology of embodiment.

In customs easily recognizable by most Roman pagans, however, incense and spices came to signify the church as sacred space and the mass as a sacred sacrifice. Sweet smells permeated the church through the use of chained censers, often made of precious metals, wherein incense or resin granules burned upon a small bed of hot coals.[86] Lamps also burned fragrant oil, suggesting a smokeless yet mysterious divine presence. At a more fundamental level, the recipes for incense and perfume included the same basic components as those found in most medicines. Thus, the church's sublime smells probably held various positive associations for their ancient audience: fragrances used to censor the mass and anoint priests, altars, and newly baptized Christians resonated with healing scents, odors prescribed to cure both mind and body.[87]

The use of incense amidst early Constantinian and Byzantine churches also mimicked an Imperial dignity available to the cult, powerfully merging sociopolitical and religious imagery.[88] Christian liturgy borrowed many representations of Imperial authority to substantiate episcopal power including the processional entry into the Church; sacred vestments available to God's anointed; and the bowing/kissing of episcopal signia (such as rings). Just as incense heralded the entry of an emperor, it also came to denote the presence of God and His hierarchy. Scent conflated the majesty of state with Church.

After examining early Christian Eucharistic rituals, a theology of embodiment—for both worshipper and Divine—appears most vividly. While the priest recreated Christ's corporal sacrifice in the Eucharistic elements, transforming bread/wine to flesh/blood, the audience smelled the mysterious fragrances of healing and hope. Church leaders enjoined Christians to become a living sacrifice in themselves, through either a glorious martyrdom or virtuous struggle against evil and vice. As the episcopal hierarchy solidified and crafted an ever-emerging orthodoxy, it provided the storehouse of precious scent largely reserved for public cult practice. Through hagiographies and sermons, they also provided corporal archetypes, perfected in Christ, for all Christians to imitate (men and women). It is to these bodies that we now turn.

SAINTLY BODIES AND SCENT

Theologians and hagiographers in late antiquity and the early Middle Ages relied upon fragrance in crafting an idealized Christian body in a variety of ways. In defining sanctity, both masculine and feminine, scent betrayed more than just outward aroma, revealing whether one had bathed; what one had eaten; or whether one was ill. Fragrance indicated, instead, a moral topography—virgin or whore? Saved or damned? Temporal or celestial? Also in shaping ideals of holiness, we find that Christian writers relied disproportionately on the female body in sacred discourse.

Early Christian male authors indeed depended upon Biblical *exempla*, martyrs, and saints when exploring what it meant to be holy. Within this rhetoric, theologians and moral polemicists tended "*to think*" with women's bodies; i.e., female figures feature prominently as standards of what *not* to do and virginity surfaces as the most significant purveyor of spiritual purity and salvation.[89] Certainly Church Fathers and hagiographers did not require chastity from all their converts and hearers; yet, they quickly pointed out the rigors of married and parental life compared to the joys of virginity. Ambrose, for example, clarifies that brides and mothers must contend with countless demands of husbands and children which brings much "weeping" and much "pain before pleasure."[90] Virgins, on the other hand, happily strive to please only God, find rest in virtue's simple beauty, and remain fertile in works of the Holy Spirit.[91]

Scent, already so meaningful in cultural context, occupies a prominent position within this discourse. Many Church Fathers identified the spiritual body—transformed through baptism and sustained through the Eucharist—with sweet smells. First, however, they unanimously condemned secular uses of perfumes among the holy (and here women receive particular rebuke). In his *Second Letter to Virgins*, Athanasius writes:

> consider other women how their senses have turned foolish so that they take delight in myrrh and pleasant fragrances that certainly are not "the sweet fragrance of Christ" (2 Cor. 2:15). It is more fitting for them to be in sackcloth and ashes so that they might preserve their virginity without danger.[92]

Ambrose of Milan boasts that while married women practice deception by applying perfume and make-up to alter their true form, only to appease their husbands and own vanity, virgins display authentic beauty through their modesty and chastity.[93] Jerome even directed holy women to avoid baths altogether as they should be embarrassed by the sight of their own nakedness; indeed, as attending to such sensual pleasures could only lead to sin.[94]

Virgins, instead, should waft the naturally sweet scents of the Spirit. Ambrose compares virginal bodies, transformed through asceticism, to the Song of Songs' "garden enclosed":

> A garden locked is my sister, my bride, a garden locked, a fountain sealed. Your channel is an orchard of pomegranates with all choicest fruits, henna with nard, nard and saffron, calamus and cinnamon, with all trees of frankincense, myrrh and aloes, with all chief spices—a garden fountain, a well of living water, and flowing streams from Lebanon.[95]

According to Ambrose, virginity blooms in the heart as a fragrant rose and naturally seeks a life of quiet solitude within a garden, "fenced in by the wall of the Spirit," instead of trampling through the impure and violent world.

The model of a "garden enclosed" effectively proscribed virgins to a modest life spent within their homes, protected from secular sin. The transformed virginal body then exudes the sweet garden smells:

> See what progress thou settest forth, O Virgin. Thy first odor is above all spices, which were used upon the burying of the Savior, and the fragrance arises from the mortified motions of the body, and the perishing of the delights of the members. The second odour, like the odour of Lebanon, exhales the incorruption of the Lord's body, the flower of virginal chastity.[96]

The physical body's transformation and immortality, secured through virginal sanctity, results in fragrant odors, imitating Christ's own burial ointment. As Ambrose says, "blessed virgins emit a fragrance through divine grace as gardens do through flowers, temples through religion, altars through the priest."[97] The virgin's dead, sinful flesh (after dying to self) resurrects to its new, incorruptible and fragrant form.

According to Ambrose and others, even though virgins might be confined within specific walls, their sanctity drifts abroad as the "perfume of holy religion" and attracts potential converts.[98] Like the blood of the martyrs, virginity's fragrance rewards not just individuals but the Church as a whole. In this they also imitate Christ, whose name provides an "ointment poured out," attracting more virgins. Ambrose relates Jesus to Song of Songs: "Your anointing oils are fragrant, your name is perfume poured out; therefore the maidens love you!" (1:2)[99]

Texts and sermons glorified virginal bodies and the sanctity they represented, but the veneration of bodies transformed by faith also took place in ritual and material culture. Expectedly, these embodied traditions reveal various olfactory conventions as Christians experienced Divine presence with their full sensorium. Described in both martyrologies and hagiographies, fragrance (sometimes both fragrant and foul) signaled the holy throughout sacred space and acts.

Beginning in the late fourth and early fifth centuries, Church authorities translated bodies of martyrs and saints into city and Church walls. Ambrose espoused the veneration of holy bodies within sacred space even though this contradicted Roman (and Jewish) pollution concerns. He certainly was not the first; the *Apostolic Constitutions* had even encouraged Christians to commemorate their dead in cemeteries "without taking precautions (for pollutions)."[100] Christian holy bodies effectively merged heaven and earth instead of posing a threat of ritual contamination: they were "contagious" in spreading saintly virtue not impurity. Because of their significance in bridging the human and Divine, they usually featured prominently within sacred space.

Indeed, the Second Council of Nicaea in 787 mandated that all altars must be sanctified by a relic, formally associating Christian identity to a *loca sancta* instead of just the Eucharistic action (as occurred earlier when Christians gathered in home-churches).[101] The *Liber Pontificalis* relates Constantine's generous contribution of scent—including incense, perfumes, and spices—to early Church monuments. We might assume from this that fragrance had appeared in earlier home gatherings; otherwise, Constantine's gift appears a novelty. Or, Constantine might have allocated Christian sacred space the same aromatics that other Roman cults employed. Whatever the case, after the fourth century, church leaders utilized saints' relics as well as scent in consecrations.[102]

Priests and church authorities generally placed corporal and other material relics throughout sacred space instead of just at the altar. In most churches, the multiple locations of relics underscored a compelling kinetic aspect of church attendance; visitors *moved* within Christian architecture and arrangement. Saints' relics might indicate an entrance or exit; or, they might collaborate in relating a story or oral tradition to a main pilgrimage attraction as a viewer progressed through the area.[103] In any case, relics sanctified altars, certainly, but their presence throughout churches indicated robust cultic activities, increasingly popular throughout late antiquity and the early Middle Ages. They also bestowed fame and prestige to the Church and local bishops/priests who administrated them.

The various cultic stations throughout sacred space might also represent a more symbolic aspect of saint veneration. Martyrs or saints were very seldom celebrated independently within visual spectacle or liturgy, which proposes a paradox of sorts. On one hand, dedication to a particular saint and prayers for intercession could be very personal and intimate. On the other, the saints, together, characterized the court of Heaven, a City of God, uniting all members of Christ's body, the Church. While a shrine might be dedicated to one saint, it also celebrated the communion of saints with multiple relics and images. At the West Church in Behyō (Syria), for example, three separate reliquaries that dispensed holy oil (in the manner mentioned below) marked the entrance/exit. The reliquaries consisted of six total openings, allowing visitors to gather oil that had contacted at least six saints.[104]

Both archeological and textual evidence provide important clues into how exactly communicants interacted with their saints. At many Syriac churches, sarcophagus-shaped reliquaries stood by entrance/exits, usually in clusters of three. These reliquaries, positioned in situ instead of movable, contained various drainage conduits that brought oil (scented?) into contact with relics. As the spiritually enhanced oil egressed the bottom holes, a valued secondary relic became available to pilgrims.[105] Less impressively, some Galic shrines had holes drilled atop the tombs' lids so that visitors whispered prayers or lowered bits of cloth to produce their own contact relics.[106]

Fragrance usually inundated these spaces, not only through censored mass and perfumed oil in lamps, but also through the saintly body's imagined presence. Saints were most notably identified, either upon discovery or attendance, by their sweet aroma. Paulinus, Bishop of Nola (d. c. 431), describes the veneration of Felix, a bishop/confessor who died c. 250. He explains that the saintly bones "breathe out the life-giving fragrance of his triumphant soul."[107] Paulinus also joyfully notes that while the "dwellers of heaven enjoy the mind of Felix, we enjoy his body."[108] Gregory of Tours includes several examples of holy bodies wafting sweet smells. At the Church of S. Venerandus, for example, one sarcophagus shattered and revealed an intact female form, with long hair, smelling of spices.[109] S. Pelagia of Limoges wished her body be displayed for four days so that "all the servants might come and see my body and none . . . might be excluded from my funeral." Before she was buried, after the requisite four days, "such a sweet fragrance flowed from her body that everyone was surprised."[110]

The saints' glorious scents usually evoke an image of Paradise, thus providing a conduit to God via the saint (believed to be present in Heaven) as well as granting a glimpse of Eternity promised to believers. Merovingian Christians responded to this powerful association by anointing or embalming their dead to reassure family and Church of the deceased's orthodoxy and virtue (as mentioned above in Part I).[111] Just as cosmetic and embalming ministrations today, these rituals probably served to alleviate grief and renew faith in resurrection and immortality.

CONCLUSION

Early Christian theology and ritual stimulated the believer's full sensorium, one transformed and renewed. The central tenet of Jesus' salvific death and resurrection, located so fully in His body, also required His followers' corporal transmutation. Through baptism, an (increasingly male) priesthood exorcised demons and buried the novitiate in blessed waters to be reborn a new creature. That creature, sealed and anointed in sweet spices, lived a new life, accustomed to a new spiritual sensorium that remained ever vigilant against the stench of immorality and sin. Renewed through the Eucharist, Christians communally ate Jesus' flesh and drank His blood in honor of His sacrifice. Many of those Christians imitated even that sacrifice through martyrdom, offering the new *ecclesia* a dramatic performance of a new Eucharist, their own bodies smelling sweet like bread and spices.

After Christianity became an established Roman cult, scent played a greater role in public ritual even though Church Fathers had earlier worried that smells such as incense would reduce Christian practice to a pagan counterfeit. In a public venue, under the growing authority of the priesthood, the correlation between Jesus' sacrifice and Eucharistic incense would not be ignored. Also, alongside tales of the martyrs' sweet-smelling sacrifice that helped identify Christian sanctity, saints, and their relics established moral exempla for believers and marked sacred space. Scent—primarily of flowers and spice—revealed the saints and their charisma, usually venerated by their bones. Priests and bishops moved these sacred remains to the center of their sacred space, aligning the sweet scents of sanctity with access to God at His (and their) altar. Churches also aligned fragrant relics along their boundaries, urging worshippers to move among the stations, allowing for greater access to the holy. Here again, as within Part I, we see space delineated by wafting aromas and, in this case, increasingly controlled by a professional clergy. A religious message that began by stressing the incorporeal nature of God as compared to demonic Roman counterparts, and the efficacy of emotional faith in a salvific Redeemer instead of the ritual fulfillment of covenant, evolved into an embodied tradition, celebrated by sound, taste, touch and, indeed, smell.

NOTES

1. See Harold W. Attridge, "Making Scents of Paul: The Background and Sense of 2 Cor. 2:14–17," in *Early Christianity and Classical Culture: Comparative Studies in Honor of Abraham J. Malherbe*, ed. John T. Fitzgerald (Leiden: Brill, 2003), 71–88. Attridge also connects scented processions with the worship of Isis and Dionysus.

2. The passage might also suggest a sense of wellness, or wholeness, available through Christ; e.g., a healing fragrance that makes the body whole, just as the triumphant warrior is unbroken. See T. W. Manson, "2 Cor. 14–17: Suggestions toward an Exegesis," in *Studia Paulina: In Honorem Johannis de Zwann Septuagenerii*, eds. J. N. Sevenster and W. C. van Unnik (Haarlem: de Ervem F. Bohn, 1953), 155–62. Manson notes that the Talmud often characterized the Torah as life-giving (healing) to some, but poison to others (unbelievers/ gentiles). Tzvi Novick also notes that Paul asserts himself as an honest, "unworldly" messenger of God as compared to "worldly" peddlers, who obfuscated with either their message or their product. He includes important connections between peddlers of fragrance as peddlers of "knowledge" present in Jewish tradition. See "Peddling Scents: Merchandise and Meaning in 2 Cor. 14–17," in *Journal of Biblical Literature* 130.3 (2011): 543–49.

3. Origen, *Dialogue with Heraclides*, 54.

4. Origen, *Dialogue with Heraclides*, 61–63.

5. Augustine, *Confessiones*, X.27, p. 232.

6. Susan Ashbrook Harvey has most eloquently traced traditions in Syriac Christianity in various works including "On Holy Stench: When the Odor of Sanctity Sickens," *SP* 35, eds. M. F. Wiles and E. J. Yarnold, 90–101 (Leuven: Peeters Press, 2001).

7. Zeno of Verona, Sermon I.32; discussed and translated in Gordon P. Jeanes, *The Day Has Come! Easter and Baptism in Zeno of Verona* (Collegeville, MN: Liturgical Press, 1995), 68.

8. See Georgia Frank's fine discussion in "'Taste and See': The Eucharist and the Eyes of Faith in the Fourth Century," *Church History* 70.4 (Dec. 2001): 619–43.

9. There are, of course, various meanings and symbolisms present in baptismal rituals. J. Albert Harrill, for example, connects the initiates' spiritual transformation as like the Roman rite of passage, the *toga virilis* ceremony. This rite publically recognizes the passage from boyhood into adulthood as he dons the white toga. In both practices, the initiate must "put on" a new maturity and accept the full responsibilities of membership in his (her) community. See "Coming of Age and Putting on Christ: The *Toga Virilis* Ceremony, Its Paraenesis, and Paul's Interpretation of Baptism in Galatians," *Novum Testamentum* 44.3 (2002): 252–77.

10. See Robin M. Jensen's fine work *Baptismal Imagery in Early Christianity: Ritual, Visual, and Theological Dimensions* (Grand Rapids, MI: Baker Academic Press, 2012). She notes that early Christian visual arts depicted baptismal candidates as children as well; see 28–30.

11. Zeno of Verona, *Tractatus* 1.24 (II.38) (*CCSL* 22:71). See also Augustine, *Sermones* 228 and 376.

12. While it seems certain that "stark naked" baptism occurred at times, Laurie Guy has argued that partial covering was more common. Instead of being "stark naked," *gymnos* or *nudus* might also connote inappropriate attire for public viewing, a form of "nudity." Thus, baptismal initiates might have worn a tunic or simple covering. This interpretation alleviates the tension between nude baptism and modesty requirements in Christian discourse. Most baptismal manuals made it clear that male priests must perform the actual water rite, even though sex-specific attendants/deacons could administer exorcism or anointing. See "'Naked' Baptism in the Early Church: The Rhetoric and the Reality," *Journal of Religious History* 27.2 (June 2003): 133–42.

13. *Apostolic Tradition* 21.3.

14. "Apostolic Constitutions" 7.1–4, in *Apostolic Tradition*, for example, explains that if "living water" (moving) is not available, still water is acceptable; if cold water is not available, warm water is acceptable; if little water is available, sprinkle it over the head three times.

15. Several studies, of course, have attempted to trace the movements of various rituals from Eastern practice to Rome, where they proliferate into general patterns (which is beyond our purpose here). For example, Elizabeth A. Leeper argues for Valentinius as the link between exorcism rites (and demonic concerns) performed first in Alexandria and then Rome. See "From Alexandria to Rome: The Valentinian Connection to the Incorporation of Exorcism as a Prebaptismal Rite," *Vigiliae Christianae* 44 (1990): 6–24.

16. Although late (late fifth or early sixth century), Dionysius the Pseudo-Areopagite approved the use of spices in chrism formulations; discussed in Turner, *Spice*, 262–64.

17. Turner, *Spice*, 262–64.

18. Zeno of Verona, Sermon I.32; Jeanes, 68.

19. See Daniel G. van Slyke's important article, "The Devil and His Pomps in Fifth-Century Carthage: Renouncing *Spectacula* with Spectacular Imagery," *Dumbarton Oaks Papers* 59 (2005): 53–72.

20. Augustine, *Enarratio in psalmum*, 50.1.

21. Augustine, *De fide et operibus*, 18.33.

22. For an important survey of various early baptismal formulae, see Peter Cramer, *Baptism and Change in the Early Middle Ages, c. 200–1150* (New York: Cambridge University Press, 1993).

23. *Apostolic Tradition* 20.8.

24. See Tertullian, *Idol*, 11.7; *Apol*, 23.16. See also Jensen's discussion, 35–36.

25. Origen, *Contra celsum*, 1.69.

26. Jensen does a fine job of linking baptism and healing in both texts and visual arts. As she notes, Church Fathers correlated the baptismal font with the Jordan River that healed Naaman the Syrian of his leprosy (see 2 Kings 5). Jesus' miraculous healings of the paralytic and deaf-mute also appear prominently in both sermons and statuary. See, particularly, 23–30.

27. Song of Songs 7:4.

28. Gregory the Great, *Liber regulae pastoralis*, 3.32.

29. Augustine, *Confessiones*, 2.3 (8).

30. Theodoret, *History of the Monks of Syria*, 1.10.

31. Theodoret, *History of the Monks of Syria*, 1.10.

32. Harvey focuses on the ethics of stench in *Scenting Salvation*, 206–10.

33. Athanasius, *The Life of Antony*, 63.

34. Athanasius, *The Life of Antony*, 64.

35. Jerome, *The Life of Saint Hilarion*, eds. Philip Schaff and Allan Menzies, *Early Church Fathers: Nicene-and-Post-Nicene, Series 2*, v. 6, p. 28.

36. William Ian Miller offers a grand examination of bodily processes, humor and culture in *The Anatomy of Disgust* (Cambridge, MA: Harvard University Press, 1997). He notes that

discussing bodily functions, like flatulence, effluvia, and foul odor became increasingly "civilized" throughout the later medieval period, yet these functions always functioned to gauge disgust or contempt. See particularly chapter 7, "Warriors, Saints, and Delicacy," 143–78.

37. Gregory of Tours, *GM*, 9. Also, see Valerie Allen's *On Farting: Language and Laughter in the Middle Ages* (New York: Palgrave Macmillan, 2007). Allen examines the meaning of flatulence throughout the later medieval period, which is beyond our purpose here. As she notes, even in the early period, not only does the "wind" represent the expulsion of demons, it might also serve to attract them.

38. Augustine, *City of God*, ed. Philip Schaff, *Early Church Fathers: Nicene-and-Post-Nicene, Series 1*, v. 2, p. 24.

39. Justin Martyr, *2 Apol* 5.4.

40. See Graham H. Twelftree's fine work *In the Name of Jesus: Exorcism among Early Christians* (Grand Rapids, MI: Baker Academic, 2007), 37–40.

41. *Apostolic Tradition* 20.8.11.

42. *Vita genovefae*, IX.44.

43. *Vita genovefae*, IX.45.

44. *Vita genovefae*, IX.51.

45. Gregory of Tours, *GC*, 9.

46. *Vita radegundis*, I.35.

47. *Vita radegundis*, I. 34.

48. 1 Corinthians 12:12–31.

49. Scholars debate how and from where these gatherings evolved. Some emphasize the meals' connection with the Jewish Passover; others emphasize their similarity with Roman banquets. Dennis E. Smith's *From Symposium to Eucharist: The Banquet in the Early Christian World* (Minneapolis: Fortress Press, 2003) explores the various components of a more general "banquet culture" of the time.

50. Biblical narratives include some of these displays of hospitality. In Luke 7:36–50, Jesus visits the Pharisee Simon's home and rebukes him for not providing water for his feet or oil for his head, both of which could be traditionally fragranced. Instead, a "woman of sin," later associated with Mary Magdeline, enters and cleanses his feet with her tears and anoints them with unguent from an alabaster jar.

51. Jonathan Schwiebert discusses the inclusion of sweet fragrance or ointment in a Coptic recension of the *Didache* and the *Apostolic Constitutions* 7.27.1–2. As he points out, it is impossible to gauge when these changes were introduced or what (if any) ritual function they played. *Knowledge and the Coming Kingdom: The Didache's Meal Ritual and Its Place in Early Christianity* (New York: T&T Clark, 2008), particularly chapter 6.

52. Novatian, *De spectabilis*, xxx. See Bowes' discussion, 19.

53. Lactantius, *Divine Institutes*, LVIII.

54. In Section 1 we briefly discussed martyr as spectacle; now, we will focus on martyrdom as a liturgical drama interpreted by its early Christian audience.

55. See Leonard L. Thompson's discussion of martyrdom and social control in "The Martyrdom of Polycarp: Death in the Roman Games," *Journal of Religion* 82.1 (2002): 27–52.

56. See David Seeley's *The Noble Death: Graeco-Roman Martyrology and Paul's Concept of Salvation* (Sheffield: Sheffield Academic Press, 1990).

57. Seneca, *Epistle* 24.7; 24.10; discussed in George Heyman's *The Power of Sacrifice: Roman and Christian Discourses in Conflict* (Washington, DC: Catholic University of America Press, 2007), 181–83.

58. Shmuel Shepkaru provides a fine review of Jewish ideas about martyrdom and resurrection in "Death to Afterlife: Martyrdom and Its Recompense," *AJS Review* 24.1 (1999): 1–44. Shepkaru argues that resurrection does not appear as a reward for sacrifice (as compared with fear of God's punishment for breaking His law) until Maccabees 4 (a later interpolation) and, most notably, in crusade literature. More importantly, martyrdom provides a way for individual sacrifice to benefit the nation of Israel as a whole. He also explores the various views of resurrection in a terrestrial plane (a new Kingdom of God) as well as celestial.

59. See Robin Darling Young's fine work in *The Procession before the World: Marytrdom as Public Liturgy in Early Christianity* (Milwaukee: Marquette University Press, 2001), 11–12.

Also, Richard Johanny, "Ignatius of Antioch," in *The Eucharist of the Early Christians* (New York: Pueblo Publishing Co., 1978), 48–70; and Finbarr G. Clancy SJ, "Imitating the Mysteries That You Celebrate: Martyrdom and Eucharist in the Early Patristic Period," in *The Great Persecution*, 106–40.

60. See Georgia Frank's discussion of vivid description in pilgrimage narratives, *The Memory of the Eyes: Pilgrims to Living Saints in Christian Late Antiquity* (Berkeley: University of California Press, 2000), particularly pp. 16–20.

61. *Ad. Rom.* 2, 4.

62. *Martyrdom of Polycarp* 15.

63. Cyprian, *Epistles* 12.2.

64. Augustine, *Sermones* 310, 2.2.

65. Ignatius, *Ad. Rom.* 6.1–2.

66. *Letters of the Churches of Lyons and Vienne*, in Eusebius, *History of the Church*, 5.1. As several scholars note (such as Daniel Boyarin and Stephanie Cobb), authors of early martyr texts masculinize female figures and in so doing appropriate many traditional, conservation gender ideals. I am not claiming here that Christian martyr texts revolutionized gender roles.

67. Oliver Nicholson provides an important look at the training of martyrs in "Preparation for Martyrdom in the Early Church," in *The Great Persecution: The Proceedings of the Fifth Patristic Conference, Maynooth, 2003,"* eds. Vincent Twomey SVD and Mark Humphries (Dublin: Four Courts Press, 2009), 61–90.

68. See Thompson, 41–43; here, he compares the training of potential martyrs with the training and oaths taken by gladiators.

69. See Andrew McGowan's insightful article "Discipline and Diet: Feeding the Martyrs in Roman Carthage," *Harvard Theological Review* 96.4 (2003): 455–76.

70. Tertullian, *Ad martyras* 1.1.

71. See particularly *On Fasting*. As McGowan suggests, Tertullian's attention to fasting and asceticism is attuned to the Montanist debate.

72. Tertullian, *Ad martyras* 1.6.

73. *Martyrdom of Perpetua* 8.1; 13.1–4.

74. Origen, *Exhortatio;* also, the most significant "general persecutions" were under Emperors Decius (r. 250), Valerian (r. 257–258), and Diocletian (r. 303–311).

75. Origen, *Exhortatio XLV*.

76. See Tertullian, *De oratione*; Origen, *De oratione*; Cyprian, *De dominica oratione; Apostolic Tradition*. Also discussed in Bowes, 53–54.

77. Gregory of Tours, *VP* 2: "Therefore the confessors of Christ, whom the time of persecution has not provoked to martyrdom, have become their own persecutors, in order to be thought worthy of God."

78. *Missale Gothicum* LXXII. *Missa unius Confessoris*.

79. See the Allen Brent's *Cyprian and Roman Carthage* (New York: Cambridge University Press, 2010), particularly chapter 6.

80. See the *Apostolic Constitutions*, 8.23.4.

81. Athanasius, *Life of Antony*, 67.

82. *Book of Pontiffs*, 34.

83. *Book of Pontiffs*, 44.

84. Leviticus 1:9.

85. As we will see in Part III, early Christian texts (canonical and apocryphal) also associated scent with Paradise, the Garden of Eden, or Heaven.

86. The clinking of a swinging censer might also be considered apotropaic, as were the sounds of bells in late antiquity. See John Chrysostom, *Homily on I Cor. 12:7*. Heather Hunter-Crawley's fine article "Embodying the Divine: The Sensational Experience of the Sixth-Century Eucharist," reminds us of the Eucharistic rite's multi-sensory appeal, including sound, smell, sight, and touch. In *Making Sense of the Past: Toward a Sensory Archaeology*, ed. Jo Day (Carbondale: Southern Illinois University Press, 2013), 160–76. Hunter-Crawley also discusses a particular fan, included in the Syrian Riah Hoarde, dating to the sixth and seventh centuries. She suggests the fan's subtle movement of air and incense simulated angelic wings (instead of

scattering flies from the bread and wine). This suggests quite a unique, and lovely sensory effect, signifying divine presence.

87. Caseau emphasizes the healing nature of scent in early Christianity see particularly her argument emphasized in chapter 4; see also Irene and Walter Jacob, eds., *The Healing Past: Pharmaceuticals in the Biblical Rabbinic World* (New York: E. J. Brill, 1993).

88. I note, here, that care must be taken while comparing early Christian and Imperial imagery. As Thomas Mathews suggests, early Church images of Jesus relate a triumphant Christ, in relation to the Greco-Roman deities. Images of Jesus did not simply "replace" Imperial representations. See *Clash of the Gods*, Princeton: Princeton University Press, 1993.

89. Peter Brown, drawing upon the work of Levi-Strauss, in *The Body and Society: Men, Women and Sexual Renunciation in Early Christianity* (New York: Columbia University Press, 1988), 153.

90. Ambrose, *Concerning Virgins*, I.VI.25.

91. Ambrose, *Concerning Virgins*, I.VI.30–31.

92. Athanasius, *Second Letter to Virgins*, 18; trans. by David Brakke in *Athanasius and the Politics of Asceticism* (Oxford: Clarendon Press, 1995).

93. Ambrose, *Concerning Virgins*, I.VI.28–29.

94. Jerome, *Letter CVII to Laeta*, II.

95. Song of Songs 4:12–14.

96. Ambrose, *Concerning Virgins* I.IX.45–46.

97. Ambrose, *Concerning Virgins* I.VII.39.

98. Ambrose, *Concerning Virgins*, I.IX.47.

99. Ambrose, *Concerning Virgins*, I.III.11.

100. *Apostolic Constitutions* 6.30.2; *ANF V. 7.*

101. Home churches in the Biblical and pre-Constantinian era would have distinguished sacrality in different ways than the architectural arrangements of post-Constantine constructions. Some of these ways included the performance of penance, the performance of the Eucharist, and proscriptions relating to behavior and responsibilities. The *Didascalia Apostolorum*, XXIII, for example, describes the need for constant attendance with much reverence and modesty.

102. Ann Marie Yasin discusses a cache of Christian materials discovered at a house church in Cirta which included a pan, perhaps used for incense, and several lamps. See her *Saints and Church Spaces in the Late Antique Mediterranean: Architecture, Cult, and Community* (Cambridge, UK: Cambridge University Press, 2009), 40.

103. Traditional scholarship has focused on the relics located within the altar, or focal point of sacred space, and then the proximity to the holy that emanates from that point (following Peter Brown's *Cult of the Saints*). Ann Marie Yasin's work, among others, instead focuses on the dispersal of sacrality and the activities of cult present within churches.

104. Yasin, 167–68.

105. See Yasin, 165–68.

106. See Gregory of Tours, *GM* 49.36.

107. *Poems of Paulinus*, ACW, v. 40, p. 181.

108. *Poems of Paulinus*, ACW, v. 40, pp. 51/102

109. Gregory of Tours, *GM* 34; this miracle is also mentioned briefly in Part I.

110. Gregory of Tours, *GM 102.*

111. See one excellent article that outlines preservation methods such as wrapping the body in linens and spices; injecting it with preservative solutions; and surrounding it with bouquets and floral wreaths. Jacques Bouillart, *Histoire de l'abbeye royale de Saint Germain des Pres* (Paris 1724); R. Boyer, "*Le sacrophage à sa découverte, in Premiers temps chrétiénes en Gaule méridionale: antìquité tardive et haut moyen âge, IIIème–VIIIème siècles, Catalogue de l'exposition* (Lyons 1986).

Chapter Six

Purifying the Body: Scent in Early Islam

Early Christians prominently included scent in their textual traditions and ritual activities. Sweet smells generally signaled transformation and transition between various categories. Rites of baptism and the Eucharist, for example, effectively transmuted the initiates into new spiritual creatures, attuned to their world with a new sensorium. Through their identification with Christ's salvific act, new Christian bodies, freed from the stench of spiritual death and decay, wafted sweet smells of sanctity often established through ascetic feats, personal charisma, and miraculous intercession. When we consider Islamic traditions, we find that early and medieval Muslims also feature sweet scents in texts and religious rituals. Just as their Christian counterparts, they utilized spices and perfumes to identify holiness and moral exempla.

Islamic views of sanctity, however, differ significantly from early Christian conversations about salvation and saints, or the *amici Dei* (friends of God). Christian emphasis on original sin (thanks particularly to Augustine) required a salvific transformation of the "old man." The Apostle Paul explained that "if anyone is in Christ, there is a new creation; everything old has been passed away; see, everything has become new!" (2 Corinthians 5:17). Muslim bodies needed no such salvific experience or ascetic heroism to confer a corporal transformation. Instead, Muhammad and later theologians celebrated corporal pleasures as those created by Allah and imagined an afterlife that included physical delights—only without threat of ritual pollution. In Islamic discourse, scent indicates holiness and sanctity, yet with more emphasis on the body's constantly shifting state of *purity*. Islamic tradition, based largely upon the reported actions of the Prophet himself, carefully routinized the body's faculties, often viewed with suspicion in Christianity. Sweat, menstruation, flatulence, and various effluvia posed complications with *tahara* (cleanliness or purification) instead of signifying demonic possession or a corrupt nature.

To understand early Islamic notions of sanctity, we must first examine not only the Qur'an, accepted by Muslims as the inimitable Word of Allah, but also the Prophet Muhammad. Muslims consider him the last and greatest Prophet as well as an archetype for moral behavior; the Qur'an refers to him as the *uswa hasana*, a "beautiful model" intended for pious imitation: "The Messenger of God is a beautiful model for those of you who put your hope in God and the Last Day and remember Him often" (Qur'an 33.21). Much like Christians with *imitatio Christi*, Muslims endeavor to imitate their beloved Prophet's honorable and virtuous behaviors, ranging from his physical appearance (he wore a beard); to his marriage practices (arranged marriage with a young virgin, `A'isha); to his prayer rhythms (times and number of prostrations).[1] Traditions about Muhammad's actions and teachings afford theological insight as well as standards for ritual performance. These are available not only in canonical hadith literature but also extensive *sira* (biography) compilations.[2] Most *sira* texts include incredible details about the Prophet's body and mannerisms; and as with early Christianity, scent permeates both text and practice, signifying purity, sanctity, and movement between various categories.

Muhammad is not the only storehouse for pious exempla, however. Throughout Islamic history, there have been models of sanctity offered alongside Muhammad and his inner-circle celebrated by hagiographers and theologians alike. Two primary groupings of piety and virtue exist: the prophets and the saints (the *awliya Allah*, or "friends of God"; s. *wali*).[3] Many of the Islamic prophets resemble those in Judaism and Christianity; distinctly, however, Muslims generally regard them as infallible models for a virtuous life (with Muhammad chief among them, of course). Such prophets include Adam, Noah, and Solomon; Muslims generally believe Jewish traditions that exposed prophets' sinful natures or destructive behavior introduced "corruption" that tainted their scripture. In Islam, prophets enjoy such an exalted status because God specifically chose them to reveal a part of His eternal message. Their immaculate lifestyles have always interested Muslims, thus we have a plethora of hagiographies about their lives and legacies called *qisas al-anbiya* (or, "Tales of the Prophets"). Their immaculate status means they never intentionally disobeyed God; as humans, however, they certainly displayed simple human frailty.[4] The Prophet Muhammad himself sometimes doubted, suffered physical ailments, and grew impatient.

Islamic traditions relating to saints (*awliya*) introduce theological complexities as well, among both Muslims and academics. Many Muslims reject the notion of saints—that is, the veneration of pious men and women who may act as intercessors before Allah. These Muslims emphasize Quranic passages such as 2.254: "You who believe, give from what We have provided for you, before the Day comes when there is no bargaining, no friendship, and *no intercession*. It is the disbelievers who are wrong." They consider the principle of intercession, or *tawassul*, particularly offensive, except for that of Allah and (perhaps) Muhammad. Other Muslims, however, recognize saintly intercession, pointing to such verses as 19.87: "no one will have power to intercede *except for those who have permission from the Lord of Mercy*" (emphasis added). According to this interpretation, saintly figures might aid believers who offer prayer and supplications only by Allah's leave.

Identifying what constitutes a formal "saint cult" or sacred biography can also be difficult for scholars because venerating the *awliya Allah* incorporates localized recognition and ritual activity. Islam has no universal sanctification "process" that rests in the hands of a few, designated authorities; instead, holy men and women generally depend upon local audience response. Even this becomes more complicated with the medieval advent and evolution of Sufi theology and brotherhoods who, indeed, regulated access to holy figures more fully while advertising them more widely.

In this chapter, I will bypass much of this argument by focusing on *how* medieval Muslims distinguished the sacred instead of arguing for the un/orthodoxy of saint veneration or assigning certain figures to "saint" categories. Early textual traditions and ritual praxis certainly included the lives and legacies of pious men and women intended for inspiration and imitation; and, as with early Christian tradition, scent distinguished those beloved by Allah. Sweet smells of perfume and flowers even signified holy bodies. Luckily, we have many *tabaqat* (sacred biographies) and *ziyara* (pilgrimage) manuals that complement Qur'an and hadith collections in revealing early Muslim conceptions of the sacred.

MUHAMMAD'S SWEET SWEAT

Perhaps the first and best example of sweet sanctity in Islam is Muhammad himself.[5] The early community went to great lengths to describe the Prophet's physical attributes as well as virtuous character. In presenting these characteristics, traditions appeal to the entire sensorium. Many of Muhammad's followers judged him to be the most beautiful of all men, and compared his visage to blazing light (or *nur*). Muslim theologians associated the *nur Muhammad* with the *nur* mentioned in Quran 5.15: "People of the Book, Our Messenger has come to make clear to you much of what you have kept hidden of the Scripture and to overlook much (you have done). A light (*nur*) has now come to you from God, and a Scripture making things clear."[6] Muhammad, in a fundamental way, provided the light of Allah, true knowledge and revelation, among his people. More mystical interpretations related the *nur Muhammad* to a primordial essence created by God before time, thus making Muhammad the first prophet and substantially distinct from other creatures. These traditions elevate Muhammad (and his descendants, according to the Shi`a) as a sublime being, intimately connected with the *nur Allah* and unique among all other prophets and Muslims alike.[7]

Traditions recounting the *nur Muhammad* range from the miraculous to the charmingly mundane. Ibn Ishaq's *Sira* describes how several women offered themselves to the Prophet's father, Abdullah, because they saw a holy glow on his forehead between his eyes. After he consummated his marriage with Amina, depositing the blazing semen into her womb, they no longer desired him.[8] As Amina carried the Prophet, she also "saw a light come forth from her by which she could see the castles of Basra in Syria."[9] The Prophet's followers often remarked on his bright light, glittering like the moon, emitting a brightness greater than celestial orbs or manufactured lamps. For Muslim hagiographers, this light proved his station, emboldened the *umma* in battle, and even aided `A'isha in locating a needle she had lost in their dark home.[10]

While Muhammad's body was beautiful to see, it was delightful to touch as well. Anas ibn Malik reported that he had never touched silk softer than the Prophet's own hand.[11] Muhammad considered visiting the sick an important responsibility for all Muslims, and he often blessed the sick with his touch or gifting his ablutions water. According to hadith transmitter al-Sa'ib, when he was a boy, his aunt took him to the Prophet for a blessing: "The Prophet touched my head with his hand and invoked Allah to bless me. He then performed ablution and I drank of the remaining water of his ablution and then stood behind his back."[12] Not only Muhammad's touch but also the water that had contacted his skin in ablutions provided the coveted *barakat* (or, blessing).

The hadith do emphasize, however, that Muhammad never touched a woman that did not "belong to him," even when she pledged her allegiance to Islam. In such cases, he simply spoke his acceptance so that "his palm never touched the palm of a woman."[13] While Muhammad appears quite modest here, avoiding the touch of women outside his kin group, he challenges some purity expectations within the marriage bond in other instances. `A'isha explains that the Prophet regularly allowed touch and even sexual play while she was menstruating, rejecting pre-Islamic and Jewish menstrual taboos. He allowed `A'isha to comb his hair; lie under the same blanket; and even fondled her above the waist.[14] He also performed prayers while sitting on the same blanket with her during her menstrual cycle.[15]

According to Muhammad, Allah only forbade intercourse with a menstruating woman; otherwise, she did not constitute a contagion. Muhammad also explains in great detail how women should purify themselves after menstruation by washing their genitalia three times with a musk-scented cloth.[16] Many women approached the Prophet and asked for his aid in identifying what type of discharge they experienced; he then judged as to whether the flow constituted menstruation or irregular bleeding depending upon color or frequency.[17] As with most patriarchal systems, traditions emphasize the male prophet's government of the female body in shockingly intimate detail, regulating the lines of purity and defining the normative (in this case menses) and the exception (erratic emissions).

While *sira* and hadith literature commend the Prophet's own body and his blessed touch, they also relate the Prophet's body with taste by regularly attributing him with food miracles. When he was an infant, his mother Amina placed Muhammad in the care of a Bedouin foster-family for suckling, a common custom among more settled Arabs. Only the poorest of the Bedu families accepted the child because he was already orphaned by his father. Halima, Muhammad's foster-mother, and her family had been suffering greatly as their pack animals withered during drought, failing to produce milk. Even Halima's breast milk dried and her own child wailed with hunger, all until the infant Muhammad came to them. With his arrival the animals began to produce milk "in abundance" even when other flocks remained barren and dry. Halima's breast milk also began to flow easily and the family enjoyed the renewed food source for two years, until Halima weaned Muhammad.[18] The young Prophet, at first scorned by Bedu families because of his orphan status, became a celebrated source of good fortune and blessings.

The Prophet as an adult also cared for his community's struggles with poverty and the dry, desert climate. Abd al-Rahman ibn Abu Bakr relates that the Prophet directed his followers to always find a place for charity: if there is food for two people, invite a third; if there is food for three, invite a fourth. On one occasion "by Allah," available food tripled for Abu Bakr's family; they fed all their guests until they were satisfied; and then they delivered leftovers to the Prophet.[19] For Abu Talhah and Sulaym's household, the Prophet's blessing increased available bread and butter.[20] Shi`ite Muslims esteem several traditions that describe Allah's special provisions for Muhammad's family through his physical presence. Centered around the Prophet's food multiplication, all these hadith emphasize the role of faith in Allah and His provision for the charitable and humble.

While hadith collections might describe Muhammad's miraculous intervention in a didactic sense, to teach faith and ethical responsibilities to others, they also reveal the *umma's* vulnerability in a harsh landscape. Muhammad proliferates food supplies as well as miraculously produces water. According to Anas ibn Malik, a man at a Friday sermon once asked the Prophet to intercede with Allah during a drought. The Prophet lifted his hands, prayed, and it immediately began to rain "heavily." By the end of the prayer session, the people waded to their homes in the water. After an entire week of such downpours, the people begged Muhammad at the next Friday sermon to ask Allah to withhold the rain as their houses began to collapse. Muhammad again prayed and the clouds parted, shielding Medina like a "crown," while the rain continued outside the city.[21] In many other traditions, water even flowed from the Prophet's fingers to provide the requisite element for prayer ablutions.[22] Muhammad's body provided a fount of blessings, both literally and figuratively.

A final element of the Prophet's exquisite body is, of course, scent. Alongside his glorious light, texture, and food production, he emitted a smell "better than musk."[23] According to one tradition, Jabir ibn Samura explained that when the Prophet patted his cheek, he sensed a fragrance as from a perfumer's scent bag.[24] Anas ibn Malik also claimed that Muhammad was sweeter than musk or ambergris.[25] His mother, Umm Sulaym, gathered the Prophet's sweat and mixed it with perfume to make a superior scent.[26] Once, after napping upon Sulaym's bed, Muhammad woke to see her gathering vials of his sweat and hair from the blanket. He asked her what she was doing and she explained that "we seek blessings for our children through it."[27] On his deathbed, Anas ibn Malik requested a remnant of the Prophet's hair be mixed with his embalming oils for blessing.[28] This demonstrates that the Prophet's sweat was not only employed in perfume production but also as a relic in itself. His physical effluvia afforded access to the holy.[29]

Umm Sulaym plays a prominent role in the various traditions regarding Muhammad's physical relics. She has a curiously intimate relationship with his body as she both procures and disperses collections of his sweat and hair.[30] These traditions certainly highlight the sacrality of Muhammad's body; his fragrance is reminiscent of the sweet smells of Paradise. The presence of Muhammad's hair in embalming oil also represents the hope of death and resurrection. Umm Sulaym's possession of Muhammad's corporal relics might also betray a more political lesson, however. While many Quraysh and tribal kin of the *muhajirun* (Muslim Meccans who travelled on the *hijra* to Medina) had familial links with the Prophet and his *barakah* (blessing), the *ansar* (Muslims of Medina) lacked such kinship claims. Umm Sulaym and her husband Abu Talhah, acclaimed for their piety among the *ansar*, instead had unique access to his physical form. This connection created a symbolic kinship where the Prophet's new family guarded (and even buried) his sweet-smelling body.[31]

While early Muslims associated the Prophet with perfume, they also reported that he enjoyed fragrance; according to Abu Sa`id al-Khudri, he once commented that he preferred musk as a woman's perfume. Al-Suyuti (d. 1505) explained that the Prophet considered musk a hot scent—which would increase the husband's desire and the chance of pregnancy.[32] The Prophet also advised his community about the appropriate use of fragrance: men should bathe, oil their hair, and perfume themselves for the Friday sermon (*khutba)* at the mosque;[33] and, both men and women could use fragrance for sexual encounters with their marriage partners. According to the Prophet, however, women particularly should be mindful of its use. He chastised women who wore perfume as they "pass by people," intending to garner attention, and labeled them adulteresses.[34] In like manner, Muhammad directed women to perform *ghusl* (or complete lustration) before attending masjid if they had earlier applied fragrance.[35] This assured they would distract no men (or women?) from their prayers. The Prophet encouraged women to wear perfume but only within their domestic domain, usually as an additional sensual pleasure during sexual intimacy.[36] This places the onus of temptation on the female but it also complements the Prophet's other policies concerning sexual modesty and humility. For example, he generally discouraged prideful adornment of gold and silks; some hair dyes (for both men and women);[37] and mutating the body with tattoos and piercings.[38]

Perhaps the best evidence of Muhammad's love for scent is his own profession: "God has made dear to me from your world women, and fragrance, and joy of my eyes in prayer."[39] This tradition reveals the Prophet's love for sensual pleasures; and these pleasures, instead of signaling the weak and sinful flesh, rank alongside his delight in prayer. The first of his earthly favorites is perhaps the most surprising. The Sunnah confirms Muhammad's love for women, always within the legal parameters of marriage, by celebrating his sexual stamina.[40] Bukhari relates that the Prophet would scent himself, visit all his wives (nine or eleven) and have intercourse with them all in one night. According to tradition, the Prophet "was given the strength of thirty men" for his conjugal rounds.[41] Praising a prophet's libido might seem strange from an early Christian perspective as their theologians linked original sin to sexual shame; however, Muhammad's prowess complemented contemporary Arabian ideals of masculinity (or *muruwwa*) which praised morality as well as strength in battle and sexual expertise. Healthy sexual desire was also considered part of human nature, or *fitra*. [42] Muhammad was a "beautiful example" in all things.

The twelfth-century Sufi master Ibn al-Arabi offered a more spiritual interpretation of Muhammad's three great loves. According to Ibn al-Arabi, the Prophet's esteem for women, fragrance, and prayer mimics a great cosmic truth also expressed as a triad: Allah as the Essence (*dhat*, a feminine noun); Adam (masculine); and women (*nisa'*, feminine). Adam, the masculine, stands between two feminine forces—God's Essence which attracts Adam/Man to Himself because he was created in the Divine's own image; and, the female, generated from man. In this triad, Adam exists as both God's servant and the master of woman; as both the passive and the active principle. As such Adam/Man worships God most fully via his relationship with woman wherein he is both active (in his love for woman) and passive (in his submission of God). Ibn al-Arabi explains that Muhammad loved women because of this cosmic completion:

> the Apostle loved women by reason of [the possibility of] perfect contemplation of the Reality in them. The best and most perfect kind is the contemplation of God in women. The greatest union is that between man and woman thus it is that Muhammad's love for women derives from the divine love.[43]

In the Prophet's declaration of his three great loves, perfume is placed in a similar position as Adam/Man in Ibn al-Arabi's cosmic triad. Perfume (*tib*, a masculine noun also associated with goodness) stands between women (feminine) and prayer (*salat*, also feminine). Perfume, when viewed as the natural aroma of women, entices man's soul into his beloved's embrace; perfume, seen as God's goodness or His own sweet-smells, pulls the soul to God, the Eternal Beloved, through prayer. Perfume maintains a kind of liminal position located between the corporal attraction to women and the spiritual attraction to God, again reminiscent of transitional states so often signaled by fragrance.

While Ibn al-`Arabi commends Muhammad's three great loves as a spiritual metaphor, early Muslim jurists also rigorously linked the physical body, its desires and functions, with sacred meaning. The Islamic attention for the body and ritual purity appears early in hadith collections and legal arguments (or *fiqh*, jurisprudence); indeed, hadith record the Prophet's own exclamation that "purity is half of faith."[44] The Qur'an lays out some basic purity expectations in 5.6:

> O believers, if you rise to pray, wash your faces and your hands up to the elbows and wipe your heads and your feet up to the ankles. If you are unclean, then cleanse yourselves; and if you are sick or on a journey, and if one of you has come from the restroom, of if you have touched women and cannot find any water, then take some clear earth and wipe your faces and hands with it. Allah does not wish to burden you, but to purify you and complete His Grace upon you, that you may be thankful.

This passage (and many others like it) requires believers to assume a state of ritual purity before encountering God in prayer.[45] From this necessity, jurists argued in infinite detail what exactly constituted "impurity" and the preferred methods of eliminating such pollution.[46] The scriptural injunction ends with a justification, explaining that God does not intend simply to make believers' lives difficult but, by the ritual actions, to demonstrate their purity as Allah's servant.

Notably, physical impurity—as jurists include such things as urination, defecation, ejaculation, menstruation, nosebleeds, and vomiting—need not elicit revulsion, dismay, or disgust. Instead, they signify simple human nature and a covenantal set of rituals that symbolically reverse the "impure" while restoring the believing Muslim to "pure." Purity rituals thus function to delineate and define Islamic communal boundaries (i.e., Muslim "insiders" against the nonbelieving "others").[47] At the same time, however, that "simple human nature" exists because of a primordial event attributed to Adam and Eve (or Hawwa' in Arabic), causing humanity's perfect paradisiacal body to decompose, decay, and die. Islamic purity rituals thus must be placed within that larger discursive context: the actions of *wudu'* and *ghusl* temporarily heal a broken body that oozes liquids, solids, and gases in its perpetual cycle of decay, rendering it pure for Allah's presence. Ritual purity provides a glimpse of the perpetual purity available in Paradise, before Adam and Eve's expulsion.[48]

Even though humanity inherits their parents' flawed form, Islamic theologians never attribute "original sin" to their offspring—there is no existential, moral blemish that passes through generations. According to *fiqh*, the body resides in a pure state until it comes into contact with or experiences pollutants. Ablution rituals *return* the body to that state; for example, if believers are uncertain whether they have infringed upon their purity status before ritual prayer (*salat*), the assumption is "no," and they need not repeat any purification rituals.[49]

Islamic jurists generally separate occasions of pollution into two categories: 1) contact with an impure substance (*najasa*), such as blood, semen, dogs, and wine; or, 2) experiencing a bodily action of major or minor defilement (*hadath*), including most emissions of the body's orifices. Many occasions of pollution relate to foul smells including excrement, urine, and flatulence. These emissions expose a sequence of death and decay that defines human existence; the body consumes food, absorbs its nutrients, and then expels the waste, all while slowly and systemically succumbing to the ravages of time. These biological systems reverse only in Paradise, when God returns humanity to its perfected form. There, bodies eat and drink without excreta, without impurity.[50]

Until that paradisiacal perfection, Muslims enter into temporary stages of im/purity through ritual ablutions with water or, in extreme circumstances, symbolic cleansing with sand. Water washes the body's extremities, from the ankles to feet and elbows to wrist. The believer also passes water over the head and orifices of the face. Neither hadith nor jurisprudence require perfume or scent for ritual cleansing before prayer. Indeed, scent is forbidden for women in public prayer; the Prophet warns that Allah will not hear any women who don perfume *before* entering a mosque.[51]

Conversely, there are no such restrictions against men wearing perfume for prayer; Muhammad instructs men to perform *ghusl* and wear their best scents for Friday sermons. Hadith also compare ablution water with perfume, if no perfume is available: "if he does not find any [fragrance], then water is his perfume."[52] Perhaps this is why donning perfume is not part of normative *wudu'* practice for males (and females) as water sufficiently prepares the body for interaction with God. Using a variety of hadith collections, many legal scholars even paralleled the washing of *wudu'* with the removal of sins.[53] In this context, the body plays a pivotal role in mapping out God's mercy and forgiveness; as the believer bathes each limb, it is purified and rededicated to Allah. In one spurious yet oft-cited hadith, Anas ibn Malik relates:

> I came to see the Prophet and there was a vessel of water before him. He said to me: "Anas, come close to me so that I can show you the proportions (*muqadir*) of *wudu'*." . . . When he washed his hands he said: "In the name of God, praise be to God there is no might and no power except with God." . . . When he rinsed out his mouth and snuffed water into his nose he said: "Oh God, instruct me in my proof and do not deprive me of the fragrance of Paradise. . . . " When he wiped his head he said: "Oh God, cover us with Your mercy and spare us Your punishment."[54]

Just as in early Christian baptism, the ritual bathing seals the body to sin and punishment. In Christian tradition, however, fragrant oil also verified the body's transformation; in Islamic tradition, water only restores the corporeal boundaries interrupted by physical (and ethical) impurities.

Muhammad's encouragement of fragrance for men as part of preparation for Friday prayers is very corporate. Muhammad admonishes men to brush their teeth (use a *siwak*) and wear perfume, *even if* they only have access to women's scent.[55] Men (unlike women) celebrate their "sweet scents" in the public Friday gathering, even enjoying the fragrance of fellow believers. Scent functions here in a very communal mode, binding the males together in public space, reminiscent of Paradise. It activates the powerful "memory" of Paradise, available to believers in prayer. Fragrance also distinguishes sacred space from mundane; Muhammad directs believers to both purify and perfume mosques when they are constructed, and then again on Fridays.[56] Even though tradition allows for prayers at home (or almost anywhere else), Muhammad recommends collective prayer for men as God then views the believer as "25 times more virtuous."[57]

The prohibitions against females wearing scent to the mosque reveal a general angst concerning women's sexuality, especially in public space. They threaten to distract not only themselves but also righteous males from prayer and contemplation. As in Christianity, texts reveal an ever-present concern for women's scent. Christian holy women (sans bath) should avoid the "false" perfumes of vain women and rely upon their sacred scent. Muslim women should don fragrance to heighten marital pleasure but must otherwise avoid its use, particularly at the mosque. Muhammad encouraged men, on the other hand, to use it at any time. Such directives, from two patriarchal systems, reveal a shared fear of female sexuality that further complicates Islamic and Christian ideals of the body and holiness.

Muhammad's allowance for men to wear even women's perfume to Friday prayer (*jum`a*) is not insignificant. The hadith generally stress repeatedly that men should look like men and women like women; wearing scent customarily reserved for females would normally be discouraged. One curious example is the use of *khaluq*, a perfume that includes saffron and other spices that can be used as dye and ointment. Tradition relates that Muhammad allowed men to dye their garments and hair with *khaluq*. According to Zayd ibn Aslam:

> [Ibn `Umar reported that]: "I saw the Messenger of Allah [SAW] dyeing his beard yellow with it, and there was no other kind of dye that was dearer to him than this. He used to dye all of his clothes with it, even his 'Imamah (turban).'"[58]

However, when the Prophet encountered men anointed with *khaluq* as scent, he required them to wash two to three times and "not wear it again."[59] Only men bound for public prayer receive a dispensation as the Prophet allows them even feminine fragrance if that alone is available.

The hadith record many personal behaviors that the Prophet discourages yet does not condemn; for example, in one collection he explains ten "dislikes" including coloring gray hair, wearing gold rings, and *coitus interuptus*. Some actions prove more dangerous to believers than others. Al-Suyuti's *al-Haba'ik fi akhbar al-mala'ik (The Arrangement of the Traditions about Angels)*, for example, describes various behaviors, usually those that affect purity status and compel angels to avoid human presence. God tasks angelic companions with recording believers' deeds (both good and bad) and protect them from evil influence; thus angels' absence renders Muslims more vulnerable and at an eschatological disadvantage.[60] Many of the occasions for "angels do not enter" traditions include those relating to foul smell: angels avoid areas where urine stagnates; during elimination in the toilet; and people with bad breath from eating garlic, onions, and leeks.[61]

Whereas angelic angst regarding effluvia fits with general rules of ritual impurity (or, *hadath*), the concern for bad breath seems more surprising. There are no dietary restrictions against consuming garlic, onions, or leeks, yet the Prophet even discourages eating them "raw" and then attending mosque.[62] He also refuses to converse with anyone who has recently eaten the foods because *he* converses with angels. He explains that "the angels are harmed by the same things as men," implying foul smells offend angels just as they do people.[63]

According to Suyuti, angels also avoid those who are "anointed with *khaluq*" until they have been washed. Here again *fiqh* scholars generally neither forbid wearing perfume nor do they classify it as a pollutant; however, custom does assign *khaluq* as a feminine fragrance. It seems angels (and the Prophet) do not approve of queer behavior. Indeed, angels also refuse the presence of men playing tambourines, a conventionally feminine ritual.[64] These regulations indicate that ritual pollution occurs not only through biological functions and contact with impurities, but also ethically suspect behavior, in this case confusing gender assignments with inappropriate scent.

Islamic traditions require ritual purity in other areas of ethical performance, including Islam's traditional "5 Pillars." These common rituals entail a statement of belief, alms-tax (*zakat*); fasting (or *sawm*) during Ramadan; making the *hajj* (or, pilgrimage to Mecca); and daily prayer (*salat*). As already discussed, scent plays a minimal role in prayer; and it is at least negligibly associated with other ritual requirements. Muhammad explains that both *zakat* and fasting/*sawm* are necessary to enter Paradise; he then compares the breath of a person fasting—considered foul by most standards—as sweeter to Allah than musk itself.[65] Scent is also present in the requisite *hajj* rites which Muhammad modeled before his death in 632 CE. Within these routines, particularly, scent (and the lack thereof) indicates the liminal translation from mundane to sacred; from ordinary to *ihram*.

Several hadith explain one of the most important moments of sacred transition in Islamic ritual—when the Muslim body passes from a "natural" state into the sacred (or *ihram*), for *hajj*, when it both looks and smells quite different from normal time. Hadith forbid several actions including shaving, cutting of the nails, sexual intercourse, and even wearing perfume upon entering *ihram*.[66] Certainly, Muslims understand Mecca and the *ka`ba* as sacred space, distinguished by sacred action, so some prohibitions such as avoiding sexual intercourse might seem obvious. Others seem less so; thus, we might ask: why are these particular actions forbidden, especially wearing perfume? If anything, based upon precedents examined above, we might expect fragrance to be required for the collective gathering of the Muslim *umma* as it anticipates paradisiacal rewards.

One possible answer is that the *hajj* recreates, in even a limited way, the sacred space of the Garden or Paradise, when human bodies existed in a pure state in perfect communion with God. According to al-Tha`alabi, the *ka`ba* even mirrors God's throne in heaven as an *axis mundi* on earth:

> Then God inspired Adam: I have a Sanctuary located directly under My Throne; so go to it and circumambulate it, as (the angels) circumambulate My Throne; and pray there, as they pray at My Throne, for there I shall answer your prayer.[67]

Herein, God directs Adam to the earthly *ka`ba* as He exiles him. All the activities forbidden during the *hajj's* ritual performance relate to the fallen body, the one inherited by Adam and Eve after their sin and exile, and practiced in a "civilized" group. For example, Ibn Kathir relates that sexual intercourse did not occur in the Garden; Gabriel taught Adam and Eve the function of their genitalia only after their exile.[68] Imitating the sexual innocence of the Garden, Islamic law forbids pilgrims any carnal intimacy. In doing so, they recreate the "natural" state of humanity, before the degradations of society and civilization. Al-Tabari also explains that all perfumes originate with the perfect Garden plants; thus, again, humanity only needed to manufacture it after the fall.[69] If anyone dons fragrance before entering *ihram*, they are required to wash their body and/or clothes.[70] In this instance, the absence of perfume or fragrance among the pilgrims instead of its presence marks Mecca as sacred space, reimagining the Garden's "naturally" sweet scent before humanity's fault.

According to many traditions, however, `A'isha perfumed Muhammad's body both before and after he entered *ihram* even though he bathed before commencing the ritual circumambulations of the *ka`ba*. This use of fragrance may signal the transition between the mundane and sacred, and then back again.[71] Other hadith suggest alternative motivations, however; `A'isha relates that after she perfumed the Prophet's body, he visited all of his wives the night before entering *ihram*. Perhaps the initial anointing simply afforded sensual pleasure during his conjugal visits; the second (at the *hajj* conclusion) marks sacred boundaries more clearly.

The Prophet Muhammad provided the *uswa hasana* for all human actions, ranging from the sacred ritual (prayer, fasting, and *hajj*) to the remarkably quotidian (using a toothbrush/*siwak*, avoiding bad breath, and using hair dye). In this case, more than any other, we see the constant slippage between im/pure and sacred/mundane so often marked by scent. Muhammad is not the only model for believers, however; medieval Muslim authors drew upon Quranic narrative as well as existing compendia of Jewish and Christian scriptural exegesis in presenting various models to imitate in purity and virtue.

PROPHETS AND SAINTS

Even though Muslims venerate Muhammad most devoutly, several other prophets provide paradigms for pious emulation. Beginning in the early period, certainly by the 900s, Muslim storytellers or *qussas* began to collect narratives that complemented the Biblical tales of prophetic heroes such as Adam, Noah, and Abraham. These materials, sometimes called *Isra'iliyat*, incorporated Jewish midrash and popular folklore. *Qisas al-anbiya*, or *Tales of the Prophets*, expounded upon often scant biblical narrative; encouraged faith; and perhaps even generated conversions. While this genre became increasingly controversial over time because of evolving political and religious authorities, they offered significant models of sanctity within early Islam and reveal a critical moment in the formation of a communal identity, effectively intersecting Islam, Judaism, and Christianity.[72] Some of the most important *Qisas* collections date to Ishaq ibn Bishr (d. 821), who was influential in Muhammad ibn Jarir al-Tabari's (d. 839) monumental *Tarikh al-rasul wa al-muluk* (History of Prophets and Kings); Ahmed ibn Muhammad al-Tha`alabi (d. c. 1035); and Muhammad ibn Abdullah al-Kisa'i (d. c. 1100).

According to many of these famous *Tales*, Muhammad presents a great paradox. He is both the final *and* the first prophet; i.e., he is born in the line of earthly prophets, starting with Adam, Noah, Abraham, and Moses, yet he is foremost in status. Al-Tha`alabi reports that God delivered the *Tales* to Muhammad so that he might learn from his predecessors' noble characteristics, and so that his Muslim *umma* might avoid the mistakes of other prophetic audiences.[73] This suggests that even though Muhammad was born last, he utilized the other prophets' lives to advantage.

Other traditions explain that Muhammad is preexistent, formed from Allah's divine *nur* or light, before creation itself. The light of Muhammad (*nur* Muhammad) resided on God's throne before He placed it in Adam's loins.[74] According to al-Kisa'i, God kneaded together white, pure soil with water of Paradise into a pearl-like form. He then submerged the pearl into all the rivers of Paradise and it unraveled into 124,000 drops, each of which formed a prophet. Therefore, "all the prophets—may the blessings of God be upon our Prophet and upon them—were created from his light."[75] Al-Kisa'i also relates that after Adam's spirit first entered his created clay form, he immediately saw the *shahada* (Islamic statement of belief, "there is no God except God, and Muhammad is his Prophet") written on the throne's pavilion.[76] Early Shi`ite traditions place the entire Holy Family (Muhammad, `Ali, Fatima, Hasan and Husayn) as divine light engraved upon the throne; not only Muhammad but also the *ahl al-bayt* (people of the house) resided with God before the first human even took shape.[77]

Al-Kisa'i conveys a tradition relating to Qur'an 7.172:

> And [remember] when your Lord brought forth from the loins of the Children of Adam their posterity and made them testify against themselves. [He said]: "Am I not your Lord?" They said: "Yes, we testify." [This] lest you should say on the Day of Resurrection: "We were in fact unaware of this."

Theologians refer to this passage, interpreted as the primordial moment when God beckons all humanity to recognize His Lordship, when explaining human nature (or *fitrah*) and the justice of Judgment Day. Humans will not be able to defend their disbelief by saying they never knew God; instead, every soul has already submitted before it even takes human form.

Many *Tales* expand upon 7.172 by describing the "future" children that God pulls from Adam's loins to examine. As might be expected, Muhammad is the first to appear and then he proclaims, "I am the first to testify to thy Oneness and to confirm my obedient service to thee."[78] After that, the "second rank" of apostles emerge and reply the same, followed by the remaining believers. Adam then blesses these beings, who stand at his right, but curses those on his left who hesitated before responding to God. On Judgment Day, Adam will recall the faces and send his progeny to Heaven or Hell, respectively.[79]

Most of the *Tales* then follow a similar pattern of prophets containing Cain and Able (after Adam and Eve), and such pious heroes as Noah, Abraham, Lot, Joseph, Moses, and Jesus. Many authors even incorporate Arab prophets, largely unrecognized by the Jews, including Thamud, Hud, Shu`ayb and Salih.[80] All of these holy men modeled virtue and loyalty to God, and many of the sites associated with their lives and deaths became pilgrimage sites. Muslims, Christians and Jews sometimes jointly visited shrines dedicated to their common prophets. The Cave of the Patriarchs in Hebron (*al-Khalil*) is perhaps the most famous; local Jews, Christians, and Muslims recognize there the gravesite for Abraham, Isaac and Jacob, as well as their wives.[81] Byzantine Christians and Muslims also venerated John the Baptist's head in the Congregational Mosque in Damascus.[82]

The evolving authority of the *`ulama'* eventually sought to systematize the veneration of charismatic prophets, especially while emphasizing that the age of prophecy had ended with Muhammad. Yet, the possibility of miraculous divine inspiration remained among holy men and women known as saints, or *awliya'* (friends of God; or *wali*, s.). One of the earliest scholars to regulate Islamic sanctity is Hakim al-Tirmidhi (d. c. 910). In his schema, saints certainly ranked lower than the prophets, partly because they were unable to legislate or deliver *wahy* (revelation). Saints instead might demonstrate *ilham* (or, inspiration) and *karamat* (miracles).[83] In addition to the historic patriarchs and matriarchs then, Islam's "living saints" provided a second category of pious exempla because of particular virtuous deeds or miracles *(karamat)*. Islamic tradition insists that only God can initiate saintly miracles, usually for individual or local purpose; this contrasts with prophet miracles, intended for nations.[84] The earliest categories of saintly figures included the Family of the Prophet (particularly descendants through `Ali and Fatima, for the Shi`a); the Prophet's Companions and Followers (especially for the Sunni); and later medieval Sufis and charismatic leaders.

Saints exude a sense of power, called *barakat*, transferred between saint, contact relic or shrine, and the devotee. Many holy men and women work miracles, and some serve as intercessors before God both before and after death.[85] Saint veneration takes many forms including supplicatory prayer (both spoken and written in letters) called *du`a'*, considered especially efficacious after pilgrimage to the saints' shrines. Critics of saint veneration, such as Ibn Taymiyya (d. c. 1328), warned that anytime pilgrims request blessing from the dead in intercession (especially for rain), they push the bounds of polytheism; they could, however, pray *for* the dead as well as *for* shrines' patrons.[86] While some *awliya'* achieve a level of piety through ascetic prowess—fasting, chastity, or perpetual prayer—it is certainly not a requirement. *Awliya'* also manifest their elect status with incorruptible bodies, exuding a sweet fragrance after death. In recognition of this, Muslims often wear perfumes or present perfumes as gifts at saints' tombs.[87]

Some early Muslims viewed this proliferation of saint veneration with skepticism. One hagiographer noted an occasion when a charlatan falsely "discovered" a holy body after a dream vision from `Ali b. Abu Talib. The deceitful itinerant had earlier disinterred the boy's body from another location and reburied it. Petitioners approached the corpse with rosewater and sweet scent; eventually, however, they discovered the false miracle when the body's odor suddenly changed, presumably becoming quite foul. The locals then punished the false ascetic publicly.[88] Pilgrimage guide al-Harawi (d. 1215) noted that several sacred sites advertised the same saint's body/part; instead of arguing over authenticity, however, he declared that "Allah knows the truth."[89] Multiple medieval Islamic shrines claimed Husayn's body, for example; Karbala, Medina, Damascus, Raqqa, and Ascalon advertised his body while Karbala, Cairo, and Najaf boasted his head.[90]

Visiting saints' shrines (*ziyara*), such as that of Husayn, becomes almost universally popular in Islam even though scriptural traditions require able-bodied Muslims to make only one pilgrimage (the *hajj*) to Mecca. While *ziyara* might lack scriptural decree, Sunnah and local tradition first provided permission and even instruction. According to a Prophet's companion, for example, *ziyara* should be made on Friday. The Prophet said: "Whoever visits a tomb before sunrise on Saturday, the dead is aware . . . because of the importance of Friday."[91] Muhammad also frequented his deceased Companions' burial places even though he cautioned against venerating the dead; "I previously prohibited you from visiting tombs, now visit them and do not say foul words (*hujr*)."[92]

Al-Ghazali (d. 1111) later advocated for visiting the dead in his pilgrimage etiquette handbook, "Book on Pilgrimage." While dedicated mainly to the *hajj*, al-Ghazali explains that, while near Mecca, it is efficacious to stop at the tombs of Muhammad, Abu Bakr, and `Umar in Medina; travel to Uhud to venerate the martyrs; and then pray at the mosque of Fatima and tomb of Ibrahim (the Prophet's son who died in infancy).[93] Al-Ghazali provides prayers to recite as well as a preferred itinerary. In his pilgrimage guide, al-Harawi references numerous shrines dedicated to the Holy Family and Shi`ite Imams along with the Companions. Medieval Baghdad, for example, advertised the bodies of Imam Musa al-Kazim and multiple Imami friends and family.[94]

Ziyara manuals and guides such as al-Harawi's, became increasingly popular throughout the medieval period even while Muslim thinkers debated practical theological challenges. Such authors struggled not so much with the actual *ziyara*—or visiting the dead. Early Christians encountered holy men and women's bodies by distributing the *bodies* via relics. Islamic purity rituals and respect for corpse integrity prevented such desecration; so, if the holy dead could not travel to pious devotees, then pious devotees would travel to the holy dead for blessing or *barakat*. Muslim scholars struggled more with the rituals performed by believers *after* their arrival.[95]

Perhaps expectedly, Jewish, Christian, and Islamic conventions intersected in their reverence for the dead. Al-Harawi even notes that in Aleppo, all three groups honored a prophet by pouring rosewater and sweet fragrance over a certain stone.[96] Both Christian and Muslim pilgrims collected (or rolled within) dust from shrines and brought fragrant unguents to rub onto tombs. For many scholars, such shared customs threatened to pollute Islam; these rituals presented innovation (*bid`a*) that might lead to polytheism (*shirk*). Jurists such as al-Suyuti (d. 1505) condemned anointing shrine walls with fragrance as well as pouring scent over graves because of their Christian ritual counterparts.[97] Ibn Taymiyya warned against practicing any festivals at locales previously esteemed by pre-Islamic pagans, Jews, or Christians.[98]

Assimilation to Jewish and Christian rituals loomed as just one source of pollution, however. Scholars also warned against the immoral atmosphere promoted by saint visitation. Jurist Ibn al-Hajj (d. 1336) emphasized the unlawful mingling of the sexes allowed during *ziyara*. Women shamefully abandoned their modesty as they interacted with male guides and other pilgrims.[99] Businessmen exploited travelers and encouraged greedy spending.[100] Al-Harawi also noted with suspicion that devotees circumambulated some shrines seven times while seeking miraculous cures; this resonated with rituals reserved for the ka`ba at Mecca.[101] This internal type of corruption of proscribed ritual and ethical norms—instead of something borrowed from the outside "other"—posed perhaps the greatest threat for most Muslim scholars.

CONCLUSION

Scent and fragrance provided a powerful, multivalent symbol for early Muslims, practitioners, and thinkers alike. In ritual praxis it can signify the shift between various states of purity; for example, menstruating women purify themselves with musk before they return to prayer and pilgrims avoid scent as they enter the *hajj*'s recreated paradisiacal state. Unlike Christian tradition, sweet scent need not signal the transmuted body, transformed from sin into Christ's likeness, found particularly in virginal perfection. Muslims value fragrance instead within marital intimacy and sensual pleasure, emulating the Prophet's own profession of love for "women, fragrance, and prayer." As with Christianity, however, Islamic tradition tasks women as the primary guardians of chastity (as well as the main sources of temptation) and warns them against fragrant seduction.

Men, on the other hand, should always wear "masculine" fragrance and are encouraged to don perfumes for corporate gatherings at the masjid. This occasion, sanctioned by an emerging orthodoxy, safely maintains the boundaries of sweet scents in public space. The more private, individual displays of piety, whether at the saints' shrine or during *ziyara*, allows for looser boundaries. Unlike the priestly supervision of saint cults in early Christianity, venerating the dead was mostly a local and audience-driven affair in early Islam. Many Muslim theologians and legal scholars came to view such rites with suspicion, and they feared the impact of unlawful innovation (*bid`a*). Saint veneration, accompanied by pleas for blessing and intercession, perhaps also allows for more private access to the holy, which was met with apprehension.

Despite legalists' concerns for *ziyara* and saint intercession, Muslims continued (down to the present time) to venerate their holy dead. From the earliest purloining of Muhammad's sweat to the commemoration of prophets and saints, Islam is "lived" through the bodies of worshipers as they make pilgrimage, offer fragrance at tombs, and pray for holy aid and intercession. As with Christian ritual praxis, scent plays an important role in demarcating both those bodies and the space reserved for the union between heaven and earth, worshipper and Divine. Yet, sweet "sacred scents" also function as a memory prompt for Christian and Muslims; it points to a time and space of corporal perfection, before Adam and Eve's sin bound the body to death and putrefaction. Fragrance not only reveals the Divine presence in the "now" but also appeals to a spiritual memory of Paradise lost, mourned, and even anticipated. We turn finally to the heavenly scents that serve, for early Christians and Muslims, as the ultimate archetype for those experienced on earth.

NOTES

1. Muslims continue to disagree about which of Muhammad's actions are exemplary (i.e., intended for all to imitate) and exceptional (i.e., granted to the Prophet uniquely). A good example is the discourse concerning arranged marriage. According to hadith, Muhammad reported: The non-virgin has more right to herself than her marriage guardian, and the virgin is to be consulted about herself, and her silence is her permission (*al-Muwaṭṭa', Kitab al-nikāh*, 331). While many Muslim jurists use this as proof against marriage without the prospective bride's permission, others point to Muhammad's arrangement with `A'ishah, arguing against required consent.

2. For a fine review of *sira* literature and some of its complexities as an historical source, see *The Biography of Muhammad*, ed. Harald Motzki (Leiden: Brill, 2000).

3. See William M. Brinner's typology in "Prophet and Saint: The Two Exemplars of Islam," in *Saints and Virtues*, ed. John Stratton Hawley (Berkeley: University of California Press, 1987), 36–51.

4. See Madelung's discussion of prophetic infallibility in his article, *`Iṣma, EI2.*

5. The seminal work on understanding Muslims' veneration of the Prophet Muhammad is Annemarie Schimmel's *And Muhammad Is His Messenger* (Chapel Hill: University of North Carolina Press, 1985). A more recent collection of fine articles is *The Cambridge Companion to Muhammad,* ed. Jonathan E. Brockopp (Cambridge, UK: Cambridge University Press, 2010). Also, see Marion Holmes Katz's fine discussion of medieval devotion to the Prophet in *Birth of the Prophet Muhammad: Devotional Piety in Sunni Islam* (Kentucky: Routledge, 2007). Katz focuses on the evolution of *mawlīd* (birthday) celebrations as a "counter canon" to the Islamic law and tradition.

6. al-Ṭabarī, *Tafsīr*, IV, 104.

7. The best introductions to the *nur Muhammad* include Uri Rubin's "Pre-existence and Light: Aspects of the Concept of the Nūr Muḥammad," *Israel Oriental Studies* 5 (1975): 62–119. These traditions play a particularly important role in Shi`ism as the Imams share in the *nur Muhammad*, making them substantially different from other creatures as well. See Mohammad Ali Amir-Moezzi, *The Divine Guide in Early Shi`ism: The Sources of Esotericism in Islam* (Albany: State University of New York Press, 1994); and Mary Thurlkill, *Chosen Among Women: Mary and Fatima in Medieval Christianity and Shi`ite Islam* (Notre Dame, IN: University of Notre Dame Press, 2007), particularly chapter 3.

8. Ibn Ishaq, *Life of Muhammad*, trans. Guillaume, 68–69. See also Ibn Kathir's *al-Sīra al-nabawiyya*, translated and published separately from his *Al-bidāya wa al-nihāya* (UK: Gardner Publishing Limited, 1988), 127–29.

9. Ibn Kathir, *al-Sīra al-nabawiyya*. Ibn Kathir includes traditions that explain that the Prophet's light shown to Basra immediately after birth, while he also miraculously dropped to his knees and gazed into heaven. *Al-Sīra al-nabawiyya*, 148. Al-Tirmidhi also describes the light that emanated from the Prophet's teeth as a signature of his beauty. See *Shamā'il al-Tirmidhi*, 14.

10. See Uri Rubin's general discussion, "Pre-existence and Light," 62–63.

11. *Bukharī* 61.70.

12. *Bukharī* 75.31.

13. *Muslim* 33.130.

14. See *Nisa'i*, 3 (The Book of Menstruation and Istihadah), especially chapter 13.

15. *Bukharī* 6.37.

16. *Bukharī* 6.12–14, 18–20.

17. See *Bukharī* 6.25; 6.31.

18. Guillaume, 69–73.

19. *Bukharī* 61.90.

20. *Bukharī* 61.87. Ibn Hajar al-Isqalānī (d. 1449) includes in his commentary on Bukharī an account where Muhammad (as an adult) multiplies milk for his followers. See *Fatḥ al-bārī bi-sharḥ ṣaḥīḥ al-bukhārī* (Beirut: Dār Iḥyā al-Turāth al-`Arabī, 1988), v. 11, 281–82.

21. *Bukharī* 61.91.

22. *Bukharī* 61.85, among many others.

23. *Bukharī* 61.62.

24. *Muslim* 43.110.

25. *Muslim* 43.111.

26. *Muslim* 43.113; see also *Bukharī* 79.55 where Umm Sulaym gathers the Prophet's sweat as well as his hair to mix with Suk.

27. *Muslim* 43.21.114.

28. *Bukharī* 79.55.

29. Al-Ṭabarī explains that the first Umayyad Caliph Mu`awiyah instructed on his deathbed that his body should be clothed in a shirt the Prophet had given him, and his eyes and mouth should be sprinkled with the Prophet's clipped fingernails that he had preserved. See David Margoliouth, "The Relics of the Prophet Muhammad," *Moslem World* 27.1 (1937): 20–27.

Medieval traveler al-Harawī also describes a mosque located between Mecca and Medina that contained particularly sweet water because the Prophet once spat in the well. Al-Harawī, 95.

30. See Brannon Wheeler's brilliant article, "Gift of the Body in Islam: The Prophet Muhammad's Camel Sacrifice and Distribution of Hair and Nails at his Farewell Pilgrimage," *Numen* 57 (2010): 341–88. Wheeler also compares Umm Sulaym's intimate relationship with the Prophet's body with Mary Magdelene's care of Jesus' body in Christianity.

31. Ibn Ishaq, *Life of Muhammad*, trans. Guilliaume, 688. Ibn Ishaq relates that Allah chose Abū Ṭalḥa to dig the Prophet's grave.

32. See van Gelder, 208.

33. *Bukharī* 11.5; 11.8; and *al-Nisa'i* 14.20.

34. *Al-Nasa'i*, 48.87.

35. *Al-Nasa'i*, 48.88.

36. See Tirmidhi, *Shamā'il* 33.219. According to this tradition, women's fragrance should be "less" than men's.

37. There are several variant traditions regarding Muhammad's views of dying the hair and beard. See G. H. A. Juynboll's "Dyeing the Hair and Beard in Early Islam: A Ḥadīth-Analytical Study," *Arabica*, Tome XXXIII (1986): 49–75; Juynboll suggests a connection between hadith supporters of dye and regional henna/perfume commerce.

38. See, for example *Sunan al-Nasa'i*, "Book of Adornment," 48. Many of these traditions relate to sexual purity and modesty, yet many others aim to separate Muslims from their neighbors more directly. For example, the Prophet reportedly used henna to color his hair but forbade dying one's hair like Jews and Christians. Tattoos were also associated with tribal slavery; Muslims were "slaves" only to Allah.

39. *Al-Nasa'i*, 36.1; *Tabaqat Ibn Sa'd*, 1.398. After his death, `A'isha proclaimed that Muhammad's favorite things in this world were food, women, and perfume; discussed in Ze'ev Maghen, *Virtues of the Flesh: Passion and Purity in Early Islamic Jurisprudence* (Boston: Brill, 2005), 16–18.

40. Muhammad's passion for his slave, Māriyya al-Qibṭiyya (the Copt), is perhaps a complex exception. According to traditions, the Prophet desired Māriyya greatly after the governor of Alexandria gifted her (and her sister) as slaves. According to pre-Islamic (and Islamic) Arabian tradition, masters had rights over slaves' bodies, which included the right to sexual intercourse. On the other hand, any offspring from such unions had legal rights of status and inheritance. Māriyya, titled the *umm walad* (or, mother of [his] child), was reportedly one of Muhammad's favorite sexual partners, causing even turmoil amongst his other wives. The early sources are quite frank about Muhammad's passionate feelings for her. See Aysha Hidayatullah's fine article, "Māriyya the Copt: Gender, Sex and Heritage in the Legacy of Muhammad's *Umm Walad,*" *Islam and Christian-Muslim Relations* 21.3 (July 2010): 221–43.

41. See *Bukharī* 5.20–21. A. Alazmeh also discusses various sources for this tradition, 223.

42. Barbara Stowasser considers sexual interest in "The Mothers of the Believers in the *Ḥadīth,*" *Muslim World* 82.1–2 (1992): 1–36. See also Maghen, *Virtues of the Flesh*, who argues for the complementary nature of Islamic purity law and sexuality.

43. Ibn al-Arabi, *Fuṣūṣ al-ḥikam*, or, *Bezels of Wisdom*, trans. by R. W. J. Austin (Mahwah, NJ: Paulist Press, 1980), 275–76.

44. See, for example, *Muslim*, 2, devoted solely to purity rituals.

45. The physical action of "facing" God in prayer is central to most rituals; see A. Neuwirth, "Face of God—Face of Man: The Significance of the Direction of Prayer in Islam," in *Self, Soul and Body in Religious Experience*, ed. A. I. Baumgarten, J. Assmann, and G. G. Stroumsa (Leiden: Brill, 1998).

46. Recent and important scholarly works on purity include Marion Katz, *Body of Text: The Emergence of the Sunni Law of Ritual Purity* (Albany: State University of New York Press, 2002); *idem*, "The Study of Islamic Ritual and the Meaning of *Wuḍū',*" *Der Islam* 82 (2005): 106–45; Brannon Wheeler, "Touching the Penis in Islamic Law," *History of Religions* 44 (2004): 89–119; Ze'ev Maghen, "Much Ado about *Wuḍū,*" *Der Islam*, 79 (1999): 205–52; *idem*, "First Blood: Purity, Edibility, and the Independence of Islamic Jurisprudence," *Der Islam* 81.1 (2004): 49–95.

47. Ritual purity's importance in defining community is explored more fully in Wheeler, *Mecca and Eden*, and Katz, *The Body of Text*.

48. We will look at the restored paradisiacal form more fully in Part III.

49. See Burton, "The Qur'an and the Islamic Practice of *Wuḍū'*," *BSOAS* 51 (1988): 23–25.

50. Al-Ṭabarī, for example, explains in *Ta'rīkh* that Adam and Eve defecated only after eating from the forbidden tree; as defecation is not allowed in the Garden, God banished them. *History of Prophets and Kings* 1: 108. See also, Wheeler, "Touching the Penis," 100–4.

51. *Al-Nisa'i*, Book 48.88.

52. *Jāmi` al-Tirmidhi*, 4.41.

53. See `Abd al-Razzāq al-Ṣan`ānī, *Muṣannaf*, 151, 152, 153–56; discussed in Katz, "The Study of Islamic Ritual," 118.

54. Muḥammad ibn Ḥibbān ibn Aḥmad Abū Ḥātim al-Tamīmī al-Bustī, *Kitāb al-majrūḥīn min al-muḥaddithīn wa'l-ḍu`af ā'wal'l-matrūkīn*, ed. Maḥūd Ibrāhim Zāyid (Aleppo, 1975), ii, 164–5; discussed in Katz, "The Study of Islamic Ritual," 137–39. The jurists' emphasis on ablution and ethics also makes sense as the purification does not correspond with the offending body part; e.g., *wuḍū'* requires washing the limbs, not the genitalia, the source of urination and defecation.

55. *Al-Nasa'i*, 14.12.

56. Ibn Mājah, *Sunan* 4.16; 4.24. Perfuming the mosque could become later Islamic Imperial displays of power. Fatimid caliphs perfumed the *minbar* (pulpit) and dome for public ceremonies, for example; see Paula Sanders, *Ritual, Politics, and the City in Fatimid Cairo* (Albany: State University of New York Press, 1994), 70–78.

57. Ibn Mājah, *Sunan* 4.53.

58. *Al-Nasa'ī*, Book 48.46.

59. *Al-Nasa'ī*, Book 48.81.

60. See S. R. Burge's fantastic work, "Impurity/Danger!" *Islamic Law and Society* 17 (2010): 320–49. Burge reimagines the threat that ritual im/purity poses to believers, including their lack of angelic presence, in contrast to Kevin Reinhart's "Impurity/No Danger," *History of Religions* 30 (1990): 1–24.

61. Al-Suyūṭī, *al-Ḥabā'ik*, §386, 567, 589, 599.

62. See *Bukharī* 10.245. This is in sharp contrast to Persian folk tradition that promotes garlic and onion as apotropaic foods used to protect not only individuals but also homes. See Peyman Matin, "Apotropaic Plants in the Persian Folk Culture," *Iran and the Caucasus* 16 (2012): 189–99.

63. See (among many other references), *Bukharī* 10.246.

64. Al-Suyūṭī, *al-Ḥabā'ik*, §590–91; see Burge's discussion, "Impurity/Danger!" 331–33.

65. *Muslim*, 13.210.

66. Saul Olyan provides insight into shaving rituals as public marker of status change, although in Judaism; see his "What Do Shaving Rites Accomplish and What Do They Signal in Biblical Ritual Contexts?" *Journal of Biblical Literature* 117.4 (1998): 611–22.

67. Al-Tha`alabi, 60. See S. R. Burge's important article, "Angels, Ritual and Sacred Space in Islam," *Comparative Islamic Studies* 5.2 (2009): 221–45.

68. See Ibn Kathīr, *Tafsīr al-qur'ān al-`aẓīm*, on Qur'an 7.22. Discussed in Brannon Wheeler's fabulous work, *Mecca and Eden: Ritual, Relics, and Territory in Islam* (Chicago: University of Chicago Press, 2006). Wheeler brilliantly links various actions that lead to physical impurity, such as touching the penis, to the recreation of Eden. He suggests that when a male touches the genitalia, he becomes conscious of those parts that were indeed absent (or unrecognized) in the Garden. Recognizing this abyss, between the perfected form and the one humanity inherited from Adam and Eve, results in the impurity that requires the symbolic act of ablution. See chapter 2 particularly.

69. Al-Ṭabarī, *Tarikh rasul wa muluk*, 1:79–87; see Wheeler, *Mecca and Eden*, 65.

70. *Muslim* 15.9.

71. *Muslim* 15.51.

72. See Jonathan Berkey's review of such controversies in *Popular Preaching and Religious Authority in the Islamic Medieval Near East* (Seattle: University of Washington Press, 2001). Also, Gordon D. Newby's "*Tafsir Isra'iliyat:* The Development of Qur'an Commentary in

Early Islam in Its Relationship to Judaeo-Christian Traditions of Scriptural Commentary," *Journal of the American Academy of Religion*, XLVII/4, Thematic Issue S (December, 1980): 685–97. Loren D. Lybarger also explores the theological and political import of such narratives in "The Demise of Adam in the *Qiṣaṣ al-Anbiyā':* The Symbolic Politics of Death and Re-Burial in the Islamic Stories of the Prophets," *Numen* 55 (2008): 497–535.

73. Al-Tha`alabī, "introduction," 4.

74. For the most succinct introduction to the *nur Muhammad*, see Uri Rubin, "Pre-existence and Light: Aspects of the Concept of *Nur Muhammad*." *Israel Oriental Studies* 5 (1975): 62–117.

75. Al-Kisa'i, 1, p. 44.

76. Al-Kisa'i, 10, p. 26.

77. See Thurlkill's discussion, *Chosen among Women: Mary and Fatima in Medieval Christianity and Shi`ite Islam*, 61–62.

78. Al-Kisa'i 29, p. 63.

79. Al-Kisa'i 29, pp. 63–64.

80. For an important discussion of Arab prophets, and the unique meaning of "prophet" in Islam, see Brannon Wheeler's "Arab Prophets of the Qur'an and Bible," *Journal of Qur'anic Studies* 8.2 (2006): 24–57.

81. Medieval pilgrim Ibn Baṭṭūṭa takes note in his *The Travels of Ibn Baṭṭūṭa*, I.

82. Ibn al-Ḥawrānī, *Ishārāt*, 16–17; Meri, *The Cult of the Saints*, 200–201. Other cities claimed John's head including Allepo.

83. See Chase Robinson's "Prophecy and Holy Men in Early Islam," in *The Cult of Saints in Late Antiquity and the Middle Ages*, eds. James Howard-Johnston and Paul Antony Hayward (Oxford: Oxford University Press, 1999), 260.

84. See Frederick Denny's discussion of sanctity in Islam. He carefully distinguishes such veneration from its Christian counterparts while also reviewing some of the arguments against "saint veneration" within Islamic tradition itself. "God's Friends:" The Sanctity of Persons in Islam," in *Sainthood: Its Manifestation in World Religions*, eds. Richard Kieckhefer and George Doherty Bond (Berkeley: University of California Press, 1988), 69–97. Also, Meri, *The Cult of Saints*, 73–74.

85. The powers exhibited by saints were certainly points of controversy. The Mu`tazilites denied that saintly persons could perform miracles; Hanafis affirmed the existence of saints and their miracles; Hanbalis denied that the dead could offer any benefit to the living and rejected rites of veneration at saints' tombs. See Josef W. Meri's discussion in "The Etiquette of Cult," *The Cult of Saints in Late Antiquity and the Middle Ages*, eds. James Howard-Johnston and Paul Antony Hayward (New York: Oxford University Press, 1999), 263–86.

86. Ibn Taymiyya, *Kitāb iqtiḍā'*, 374, p. 293. See M. U. Memon, *Ibn Taymīya's Struggle against Popular Tradition; with an Annotated Translation of his Kitāb iqtiḍā' al-ṣirāṭ al-mustaqīm mukhālafat ahl al-jaḥīm* (Paris: Mouton & Co., 1976).

87. See Valerie J. Hoffman's discussion in "Muslim Sainthood, Women, and the Legend of Sayyida Nafisa," in *Women Saints in World Religions*, ed. Arvind Sharma (Albany: State University of New York Press, 2000), 107–44.

88. Sibṭ Ibn al-Jawzī, *Mir'āt al-zamān fī ta'rīkh al-ayān*, 2 pts. (Hyderabad, 1951–1952), 8(1): 176; discussed in Meri, *The Cult of Saints*, 78–79.

89. See al-Harawī, *Kitāb al-ishārāt*, 19; one example of multiple sites from Tiberias.

90. See Meri's discussion in "The Etiquette of Cult," 266; and Khalid Sindawi, "The Head of Husayn ibn `Ali: Its Various Places of Burial and the Miracles That It Performed," *Holy Places in the Israeli-Palestinian Conflict*, eds. Marshall J. Breger, Yitzhak Reiter, and Leonard Hammer (New York: Routledge, 2010): 264–73.

91. Al-Ghazālī, *Iḥyā `Ulūm al-Dīn*, 4 vols. (Cairo, 1955), 4, 491.

92. Al-Ghazālī, *Iḥyā `Ulūm al-Dīn*, v. 4, 490, 492.

93. Al-Ghazālī, *Iḥyā `Ulūm al-Dīn*, v. 4; throughout "Book on the Secrets of Pilgrimage."

94. Al-Harawī, *Kitāb al-Ishārāt*, 72.

95. Christopher Taylor provides a fine overview of these arguments in chapters 5 and 6, *In the Vicinity of the Righteous: Ziyāra and the Veneration of Muslim Saints in Late Medieval Egypt* (Leiden: Brill, 1999).

96. Al-Harawī, *Kitāb al-ishārāt ilā ma`rifat al-ziyārāt*, 4. For a lovely survey of modern shared sacred space between, particularly, Palestinians and Hindus, see Margaret Cormack, *Muslims and Others in Sacred Space* (Oxford: Oxford University Press, 2013).

97. See al-Suyūṭī, *al-Amr bi-al-ittibā wa al-nahy `an al-ibtidā`*, ed. M. H. Salmān (al-Dammām, 1990), 115; Meri, *The Cult of the Saints*, 129.

98. Ibn Taymiyya, *Kitāb iqtiḍā'*, 190, p. 198.

99. Ibn al-Ḥājj, *al-Madkhal*, 1: 267–68. For an important modern look at pilgrimage and the creation of women's space see Donna Honarpisheh, "Women in Pilgrimage: Senses, Places, Embodiment, and Agency. Experiencing *Ziyarat* in Shiraz," *Journal of Shi`a Islamic Studies* VI.4 (2013): 383–410.

100. See Meri, *The Cult of the Saints*, 127–28; M. Fierro, "The Treatises against Innovation (*Kutub al-Bida`*)," *Der Islam*, 69 (1992): 204–46; Taylor, 58–63.

101. Al-Harawī, *Kitāb al-ishārāt ilā ma`rifat al-ziyārāt*, 6.

Part III

Scents of Paradise

After examining the variety of ways scent signals liminality—movement among sacred time and sacred space—it seems fitting that we find ourselves ultimately in Paradise. In Christianity and Islam, we have discussed how aromatics helped identify both sacred and sin, righteous and wretch. Ritual practice incorporated sweet smells in purifying the body and sanctifying space. Yet scent also hearkens to the primordial memory of Paradise—perhaps the best example of liminality we have seen. Paradise exists in the past and in the future; on earth and in heaven; in basic human nature and hopeful expectation. Likewise, scent excites the body, here and now, while still recalling a loss—a painful exile and an anticipated return to the Divine. In this chapter, we will focus finally on Paradise as the ultimate sensory encounter, appealing to pious believers' imaginations and modeling ritual action. First, however, we will consider what gardens signified within early Christian and Islamic cultural milieus.

Within Christianity and Islam, Paradise is a garden; the term itself stems from a Median term meaning, *pari* (around) and *daeza* (wall).[1] The word referred to a variety of enclosed spaces including hunting parks or food storage compounds. Most commonly, *paradeisoi* implied gardens complete with trees and a variety of fragrant plants. In Near Eastern and Jewish texts, paradise/garden (both real and symbolic) usually suggested royalty, with kings serving as sublime gardeners; in Greco-Roman culture, they indicated elite status. In Arabian and Persian contexts, gardens symbolized civilization amidst an otherwise uncultivated landscape, both literally and figuratively (e.g., the presence of art, music, and poetry). As enclosed spaces they also suggested a division between public/exterior and private/interior space that helped reify social customs and gender expectations.

In perhaps all cultures, gardens signify a time and space beyond themselves; life and death revolve in a delicate cycle, prompting fond memory of the past/seedlings and hope for a verdant future/fruition. Space negotiates between violent acts of cultivation and peaceful serenity amid civilization. Roman author Columella (d. c. 70 CE) compares tilling the soil with lashing a slave and ploughing as disemboweling the earth.[2] A time and space set apart, gardens aim for a harmonious balance with nature while still allowing for transformation, transgression, and perfection. One extant garden wall's inscription in Pompeii exhibited three rules that provide a glimpse of the owner's idealized space:

> The server shall wash and dry the guest's feet; a napkin shall protect the cushions and care shall be taken with the linen.
> Cast not lustful glances and make not eyes at another man's wife, be chaste in speech.
> Refrain from anger and insolent language, if you can if not return to your own house.[3]

These garden regulations identified the cultured, urbane expectations of visitors, curtailing any unwanted passions that might disturb the carefully nurtured serenity.

Many Romans posted Priapus statues as protectors of bucolic ideals. The ithyphallic figurine complemented the plot's promised fertility but also threatened any thieves with anal rape: "When desire for the fig overcomes you and you wish to reach a hand this way, look upon me, thief, and think about what a heavy cock you will have to shit back out."[4] The garden, just as an enclosed womb, might be inappropriately penetrated because of greed or desire, but Priapus swore a swift retribution for those who dared disrupt its peace.

During the Roman Republic and Empire, garden landscapes flourished in both private and public venues. Pompey the Great (d. 48 BCE) commissioned the first community park, the *Porticus Pompeiana*, including fragrant flowers, trees, water features, and even a theater. Such grounds certainly advertised great men's political and military prowess, yet Roman gardens, popular across social strata, served as more than just propaganda pieces. The fact that the Roman city of Pompeii consigned roughly 17 percent of the town's land to gardens demonstrates the wider social significance of cultivated space.[5]

Drawing upon their Greek predecessors, for example, many Romans withdrew to their gardens in philosophical retreat from the world. Inspired by philosopher Epicurus (d. 270 BCE), Romans embraced nature and humanity's place within it. From his home and garden school, Epicurus had originally extolled a life of political and social retirement, resulting in the idealized *ataraxy*, a sublime calmness devoid of mental anguish. Roman poet Lucretius (d. 55 BCE) later elaborated on this philosophy and explored nature's intricate workings and humanity's material relation to it (i.e., relating humanity to laws of nature instead of divinity). In his *De Rerum Natura*, he equates ignorance of nature with bondage and fear: "This terror then, this darkness of the mind, / Not sunrise with its flaring spokes of light / Nor glittering arrows of morning can disperse / But only nature's aspect and her law (2.58–61)." Retreating to the garden to celebrate universal truths available to reasonable men provides for a happy life, one of *otium* (or, leisure).

Roman gardens also represented a more basic, nostalgic longing for a simple and morally superior past. This resonated with various Roman philosophies beyond those of the Epicureans. Roman scholar Varro (d. 27 BCE) recalled an uncomplicated time when humanity coexisted with nature in harmony. This impulse did more than differentiate the Romans from their classical Greek predecessors who linked civic responsibility with virtue; it hearkened to a primitive impulse, a return to nature. According to Varro's *On Agriculture*, three general rural economies existed: farming, animal husbandry, and pasturage. He ranked farming—land cultivation—as both the most ancient and the most virtuous, praising Romans who returned from the city to the country (3.1.4). Columella also lamented the loss of manly agriculturalists to the city's effeminate ways, including theaters and games.[6] Honorable Romans had forsaken Romulus' styles of hunting, toiling, and hard "labours of peace."[7]

By the Imperial era, many Romans struck a compromise and brought the country gardens within the Roman villas by incorporating peristyle courtyards into their architectural schema. Filled with fragrant flowers, water founts, statues and altars, the enclosed Paradise promised rest, retreat from the world, and a locale for religious ritual performances. The space reconciled wild, untamed nature and the public expectations of city. Framed windows provided the household cherished panoramic views; when that was not an option, landscape frescoes presented idyllic and even sacred scenes.[8]

As the Empire expanded in geography and power, gardens afforded a material glimpse of Rome's domesticating power. Columella's agricultural treatise lists the best seeds to plant at what times of the year. The great variety of species—from Spain, North Africa, Gaul, Asia, and Greece—furnished a microcosm of Roman rule. The well-tended garden, stripped of riotous weeds, mirrored the idealized Empire. Within their gardens, a space fecund with transformation and change, Roman moralists remembered their virtuous past and envisioned an eternal Empire blessed by the gods. Christian discourse readily adapted these cultural cues as they imaged an eternal paradise, complete with banquets, peace, and God's eternal presence.

Islamic gardens prove more difficult to characterize because of the multiplicity of climates and cultures within the medieval Muslim empires.[9] As with the Romans, gardens functioned within agricultural, urban and private landscapes. Yet Muslim communities from Spain to Persia identified with some common Arabian cultural roots and popular lore that they absorbed into their own cultures, relating paradisiacal traditions to material realities from the Arabian Hijaz. Pre-Islamic poetry springing from the desert relied heavily upon natural imagery in expressing love and loss, power and desolation. Early Islamic literature inherited these styles, identifying nature as an *aya* (or, sign) that reveals God's power. Contemplating the earthly garden as a window to Heaven inspired much Islamic poetry, especially among Sufis.

Islamic garden archetypes across the Empire incorporated shade and water more vigorously than their Roman counterparts, perhaps due to these desert ideals. Indeed, the Qur'an 77.41–43 teaches: "The righteous will be among shades and springs, and fruits from whatever they desire, [being told] 'Eat and drink in satisfaction for what you used to do.'" In order to create a space with water and shade, then, Muslims adopted and expanded upon irrigation techniques practiced in Rome and Persia. This technology allowed for the "greening" of previously desolate lands and then the transference of new species of plants and various botanicals.[10] Numerous almanacs, botanical treatises, and instruction manuals for practical agricultural strategies proliferated under the Umayyads and Abbasids.

The most basic structure of Islamic Gardens reflects Persian influence. Termed the *chabar bagh*, it includes an enclosed courtyard with four water conduits along cross-axial lines. In pre-Islamic Persian contexts this design usually correlated gardens as quadrants, with or without water channels.[11] In the Islamic context, the four channels provided a functional water source for the gardens' vast flora; in some cases, they also served as a symbolic reference to the four paradisiacal rivers.[12] The earliest gardens adorned palaces, then mosques, and then cemetery grounds. They generally eschewed vibrant images and statues and instead relied upon trellises, perfumes, fountains, and textiles for aesthetic grandeur.

For Muslims, paradisiacal gardens described in the Qur'an and tradition provided the ultimate archetype for earthly groves. For example, the Qur'an depicts Prophet Solomon's palace grounds when the Queen of Sheba visits. Beginning in 27.44, Sheba wonders at the gleaming floor and lifted her skirts because she perceived it as water. Likewise Umayyad Caliph al-Ma'mun (r. 1043–75) had installed a crystalline floor with water flowing from a central fountain. When visitors encountered the pool, unable to distinguish between water and crystal, they surely remembered Solomon.[13] Other rulers transformed their gardens into a kind of theater, draping trees with textiles and jewels to imitate the trees of gold, with silver and precious gems in Paradise. They might also plant multitudes of fragrant fruit trees and incorporate doves and nightingales for full sensory fulfillment. Such glorious spectacles certainly announced wealth, power, and piety; as such, they invited criticisms of worldly decadence and self-indulgence.

Garden grandeur culminated perhaps with the addition of moving forms—animated by hydraulics—that became fashionable with the elite.[14] Figures, called *automata,* might include all types of animals (such as peacocks) or more abstract designs. One of the first instructional texts in Arabic is the ninth-century *kitab al-hiyal* (*Book of Ingenious Devices*) by three brothers called the Banu Musa bin Shakir. Such mechanical wonders included not only figural water spouts but moving birds that sang. Later thirteenth-century scholar al-Jaziri compiled his own *Book of Ingenious Mechanical Devices* featuring clocks, hydraulics-driven drink dispensers, and medical tools that inspires engineers even today. All of these sensory wonders functioned together in the famed pleasure gardens of early medieval Islam.

NOTES

1. See Jan N. Bremmer's detailed study "Paradise: From Persia, Via Greece, into the *Septuagint,* " in *Paradise Interpreted: Representations of Biblical Paradise in Judaism and Christianity*, ed. Gerard P. Luttikhuizen (Leiden: Brill, 1999), 1–20.
2. Columella, *De rustica*, trans. E. S. Forster and Edward H. Heffner, *On Agriculture*, LOEB Classical Library (Cambridge, MA: Harvard University Press, 1941/1955) 10.67–73; see Pagan's discussion, 20–21.
3. See Victoria Emma Pagan's discussion in *Rome and the Literature of Gardens* (London: Duckwork, 2006), 11.
4. W. H. Parker, *Priapea: Poems for a Phallic God* (London: Croom Helm Ltd., 1988), 69; also discussed in Pagan, 24.
5. Annette Giesecke, *The Epic City: Urbanism, Utopia, and the Garden in Ancient Greece and Rome* (Cambridge, MA: Harvard University Press, 2007), discusses Pompeii, 102–3. Herein, Giesecke draws upon the work of M. Conan, "Nature into Art: Gardens and Landscapes in the Everyday Life of Ancient Rome," *Journal of Garden History* 6.4 (1986): 348–56.
6. Columella, *De rustica*, 1.15.
7. Columella, *De rustica*, 1.16–17.
8. Geisecke provides a brief discussion of framed landscapes, both natural and fabricated, 113–16.
9. Also, it should be noted that here, our purposes deal with religious and cultural *meaning* of Islamic gardens. As Fairchild Ruggles quite rightly points out, to fully understand garden landscapes, one must account for landscape, agriculture, and water supply instead of focusing only on religion and politics. See *Islamic Gardens and Landscapes* (Philadelphia: University of Pennsylvania Press, 2011).
10. See particularly Ruggles, *Islamic Gardens*, 25–27.
11. Ruggles, *Islamic Gardens*, 39–40.
12. Ruggles warns against applying a blanket generalization too early, and too often that the four water conduits always signified the rivers of paradise; see *Islamic Gardens*, chapters 7 and 8 particularly.
13. Fairchild Ruggles, *Gardens, Landscape, and Vision in the Palaces of Islamic Spain* (Pittsburgh: University of Pennsylvania Press, 2011), 147–48.

14. Byzantine palatial gardens boasted their own *automata*, certainly sharing in some common Greek inspirational sources. See Gerard Brett's "The *Automata* in the Byzantine 'Throne of Solomon,'" *Speculum* XXIX.3 (1954): 477–87.

Chapter Seven

Heavenly Bodies in Christianity and Islam

According to Christian theology, both damnation and redemption occurred in a garden: Adam and Eve disobeyed God in Eden, resulting in humanity's sinful nature (Genesis 2–3); and, Jesus' crucified body resurrected in a garden tomb and Mary Magdalene later confused Him for a gardener (John 19:38–42; 20:15). Jesus' salvific act through crucifixion effectively reverses the punishment ascribed to Adam, keeper of another garden. While early Christian theologians imagined humanity's return to that garden/paradise in a variety of ways, they all include the Christian body transformed with a new sensorium, different yet familiar.

Locating paradise and the Garden of Eden in late antiquity proves a difficult task; canonical texts describe a lush abode filled with water and precious gems (Ezekiel 28:12–14; 47:12; and Revelations 21:11–22). Its exact position varied, however; the sacred place generally shifted between earth, the heavens, and even the interior/"heart." Several Jewish pseudepigraphal texts, such as *Enoch 1–3* and *Ezra* (dating approximately between third century BCE and fourth century CE), describe paradise both as the point of creation and an eschatological destination, while situating it on earth and among the three or seven levels of heaven.[1] In *1 Enoch*, the angel Gabriel took the visionary Enoch on a tour of Earth, including the Garden of Eden in the northwest which held God's throne. That throne, surrounded by fragrant trees, provided God a resting place during his visits to earth; the trees also supplied a delicious fruit destined for the elect after judgment. One particular Tree of Life granted "long life" to the righteous as its fragrance would become part of their "bones." In comparison, the "accursed" would reside in a barren, hard landscape, devoid of natural beauty and without sensory appeal.[2]

The exact location of Eden remains unstable, however; Gabriel later journeyed east and revealed to Enoch the "garden of righteousness" where "your precursors ate, came to know wisdom and were expelled." Again, sweet fragrance marked the space, with trees redolent with cinnamon, nard, and pepper.[3] The Tree of Wisdom stood as the most beautiful and fragrant of them all.[4]

The pseudepigraphal *Life of Adam and Eve* (or, the *Apocalypse of Moses*, dating perhaps to the first century CE) mentions both an earthly paradise as well as a celestial abode of God. The text elaborates on humanity's first sin, placing the fault directly on Eve. It even begins by describing her second error, incurred because she could not perform her penance correctly. Adam had proscribed their penance after their expulsion from paradise. He required them to beseech God for forgiveness while standing in running water. Adam pledged to remain in the Jordan River forty days; he directed Eve to stand in the Tigris for thirty-seven days (because she could not "do so much"). Satan transformed himself into the "brightness of an angel" and tricked her into leaving before she had completed her atonement.[5] Grieved by her ignorance and culpability, she fled from Adam and remained in solitude until she felt labor pains. She then gave birth to Cain, reunited with Adam, and they had thirty additional sons and daughters.[6]

This text emphasizes God's mercy in finally forgiving Adam's sin as he later accepts him back into paradise upon his death. Scent and sweet fragrance signal this return by marking Adam's repentance as well as the sacred space itself. According to the text, Adam requested fragrant herbs from paradise when exiled "out" on the earth; God relented and commanded the angels to provide him with crocus, nard, calamus, and cinnamon. This request complements the Islamic traditions that describe Adam's expulsion from Paradise along with fragrant plants; those narratives relate his heavenly souvenir to his grief (i.e., he longed to remember paradise with their sweet scents). The *Life of Adam and Eve*, on the other hand, stresses Adam's sinful act and need for absolution. Adam explained that he would use the sweet fragrance as sin-offering to God, thus assuring that God could "hear me."[7]

After a long life filled with sin offerings and penance, Adam died and his soul departed his body, ascendeding into the seven heavens. As the angels veiled the firmaments with censers and incense, God pardoned Adam and ordered the angels to return him to Paradise (the third heaven) to await the final resurrection. God then allowed Paradise's sweet scent to waft out upon humanity and all (save Seth) fell asleep. The angels prepared Adam's body (as well as Cain's) with linen and fragrant oil, and allowed only Seth to know the burial place. God then promised Adam (and the audience) a final resurrection for the soul's reunion with the body.[8] The narrative mentions at least three distinct sacred places altogether: the earthly paradise, now guarded by an angel with a flaming sword; the third level of heaven's paradise, where souls (like Adam) find rest; and, finally, God's abode, located in the seventh heaven.

2 Enoch and *3 Enoch* also suggest an earthly Garden of Eden that mysteriously connects to its heavenly prototype. In *2 Enoch*, the visionary describes Eden with an open sky that reaches to the third heaven (out of the seven) where Paradise resides with a "rest . . . for the righteous" and is "indescribable for pleasantness and fine fragrance."[9] Here, between the corruptible and incorruptible sets of celestial spheres, two streams emerged, one with milk and honey and the second with oil and wine.[10] When Enoch reached the seventh heaven, he saw God who immediately directed Archangel Michael to strip him of his garments and "anoint him with my delightful oil" and myrrh.[11] Another angel then brought Enoch several books already scented with myrrh and a pen so that the prophet might record his lessons.

While *3 Enoch* also describes the Garden of Eden's link with a celestial Paradise, it relies upon visual stimuli in describing the sacred places. Instead of sweet smells signaling God's presence, His Shekinah or holy glow announces sacrality. Of all the sensory cues present in *3 Enoch*, indeed, voice and sight dominate the text. Blinding light heralded the divine Shekinah and Metatron's voice overwhelmed the visionary, Ishmael (Metatron is the transformed, now angelic Enoch). This might relate to *3 Enoch*'s function as *merkabah* or chariot literature, emphasizing the ecstatic ascent through the seven levels of heaven to God's throne. Instead of defining earthly and paradisiacal geography, *Enoch 3* offers a profound model for spiritual imitation. Mystics also sought the face of God, apparently offered to few and fraught with danger; indeed, Ishmael gazed upon "the right hand of the Omnipresent One" where even the "seraphim and ophanim were not allowed to look."[12] The divine Sound and Light successfully affirm God's inherently ineffable quality. Ishmael at first shrank from the Shekinah's sheer radiance and then approached God's right hand at Metatron's insistence, the hand that "no mouth can tell its praise, no eye behold it, because of the magnitude of its greatness, its praise, its glory, its honor, and its beauty."[13] Metatron's voice and God's light resist the human sensorium in defining the undefinable.

The apostle Paul describes a similar ascent into the heavens, where he (or a companion?) encounters the Divine.[14] According to 2 Corinthians 12:2–4,

> I know a person in Christ who fourteen years ago was caught up to the third heaven—whether in the body or out of the body I do not know; God knows. And I know that such a person whether in the body or out of the body I do not know; God knows—was caught up into Paradise and heard things that are not to be told, that no mortal is permitted to repeat.

In this description, Paul relegates the body to an insignificant, even unknown variable; was he even in his corporal form? It does not matter. Spiritual union and divine knowledge in Paradise transcend the body. He also suggests that such unity should not even be discussed, questioning the ability for regular mortals to understand.

Disregarding the "do not repeat" warning label, a popular late antique text, *The Apocalypse of Paul*, narrated Paul's visionary experience. The first chapters emphasize justice; Paul sees a righteous man and an impious man die and then their spirits/souls depart the body (although they will return there on the day of resurrection). The spirits/souls appeared before God and received their judgment. The angel Michael led the pious man to a "Paradise of joy" to join "all the saints"; the evil man not only reviewed his sins but confronted his victims. The angel Tartarus finally delivered the damned man to the "lower prison" until the day of judgment.[15]

Paul then visited the same places that the angels recently had commissioned the souls. To the paradisiacal third heaven stood golden doors and columns, allowing only those with "goodness and purity of body."[16] Even though Paul remained mostly indifferent to physical ascent in his letter to the Corinthians, it becomes much more important in this narrative. Like Paul the righteous souls themselves were "out of body"—they would rejoin their corporal form only at the resurrection. Yet, entrance into Paradise requires "bodily purity" and the joys therein appeal to the corporal senses. Paul saw rivers of milk, honey, wine, and oil that offered abundance, growth, refreshment, and rest to the virtuous.[17] The body effectively mediates the visionary's experience not only for Paul himself but also for the early Christian audience.

When Paul traveled to the place of torment, he crossed beyond an ocean surrounding the earth into a space devoid of light and filled with sorrow. Again the body conveys the depth of despair that awaits unrepentant sinners. Expectedly, Paul confronted the greedy, magicians, adulterers, and fornicators. Quite unexpectedly, perhaps, the apostle observed a plethora of Church officials who had failed in their vocation: bishops who never pitied widows stood in fire up to the knees; extorting deacons suffered bloody hands with worms crawling from mouth and nostrils; and, fornicating presbyters who illicitly served at the altar agonized over three piercing hooks in their bowels. Even those who "broke their fast before the appointed hour" desperately stared at water and fruits that hung forever out of reach.[18] Such vivid imagery extended not only a warning to sinners in general but also a powerful critique of Church officials.

The author of the *Apocalypse of Paul* probably knew a related text dated a bit earlier, the *Apocalypse of Peter*, perhaps the earliest Christian text that tours the afterlife's pleasures and punishments. In details reminiscent of the martyr stories, the author evokes reader/hearer response by appealing to all the senses even more powerfully than pseudo-Paul. According to the text, Jesus revealed the "souls of all people" to Peter and their impending rewards and tortures. First, sinners met appropriate ends as their sentences matched their crimes. Blasphemers hanged by their tongues; fornicating women dangled from their braided hair and men from their loins; and child killers resided in pits of excrement.[19] The elect, including the "righteous fathers" Abraham, Isaac, and Jacob, enjoyed better accommodations. The virtuous rested in a great garden, filled with trees, fruits, and fragrance.[20] The author engages the body's entire sensorium identifying the heavenly realms.

This heavenly, eschatological Paradise plays a much more significant role in early Christian rhetoric than Jewish tradition. Christian soteriology relies upon Jesus' garden resurrection to reverse Adam and Eve's initial crimes (concisely defined in Romans 5:12–21). Both of these sublime gardens—the earthly Eden and celestial Paradise—appear quite prominently in S. Augustine's writings. Augustine's *Confessions* recount his childhood, philosophical studies, early professional life as a rhetorician, and conversion. Augustine begins Book 8—the book wherein he becomes a Christian—with a story of Victorinus, a great teacher and philosopher. According to Augustine's friend Simplicianus, Victorinus identified himself as a Christian secretly yet feared public ridicule from his elite friends. Victorinus struggled to make the private, public—yet finally did so by making his profession before the Church members. Augustine's own struggle existed still in the private "will" of his heart; he studied the Bible, pondered Paul's writings, yet resisted the pull of his "flesh," particularly the lust he felt for women.

According to the *Confessions*, after his colleague Ponticianus related the story of Egyptian hermit S. Anthony and described the abundant monastic life followed by many Christians, Augustine experienced a profound anguish. Perhaps inspired by the story of eremitic withdrawal and solitude, Augustine retreats to his garden and cries out to God. He loathes the way his will—split between both carnal and spiritual goals—struggles against itself; contrary to his Manichaean background, he recognizes this as humanity's inheritance from Adam's original sin, not as two substantive, warring selves (one good and the other evil). He then hears a child singing the background, "pick it up, and read it."[21] He finds a Bible, opens it and reads the first passage he sees just as S. Anthony did when he accepted his monastic vocation (Matthew 19:21). Augustine found a Pauline passage: "Not in rioting and drunkenness, not in chambering and wantonness, not in strife and envying, but put on the Lord Jesus Christ, and make no provision for the flesh to fulfill the lusts thereof (Romans 13:13).[22] Augustine's weeping ceased and God transformed his divided will into one, without fear or doubt. He found his friend Alypius who experienced his own conversion and then sought out his mother, Monica, to share his good news.

Augustine's conversion in a garden serves a significant theological and didactic purpose. When Adam and Eve originally sinned in the Garden of Eden, they first recognized their nakedness—their sexuality—and covered themselves. Augustine equates this initial sexual enlightenment with lust, what mars the originally perfect and pure humanity. When he sits, weeping in his garden, he sees himself (Adam's descendant) at first helplessly bound by carnal desires—lust surely but also greed and pride. God transforms that flesh, however, and returns him to the prelapsarian hope for perfection; Augustine then "sought neither a wife nor any of this world's hopes" and informs happy Monica that she should no longer hope for "grandchildren of his flesh."[23] God transmutes Augustine's body—both spiritual and carnal—in the garden space, located between his past life and his hope of salvation; his untamed, sinful self and the perfected body promised in the resurrection.

Augustine defines the connections between terrestrial and celestial Paradise most fully in his *City of God.* There, Augustine outlines the earthly city, composed of Christians living among sinners in this world, and its eschatological transformation into the heavenly city after the resurrection. Augustine quite forcefully insists that both cities exist in "reality," and do not merely function as allegory.[24] The first humans—contented and immortal in the Garden of Eden—transgressed God's law and thus planted the "seed of death," transmitted to every human being through sexual intercourse.[25] That initial offense caused the body/flesh to lust against the spirit, with God's punishment being mortality.

While Augustine identified sin's initial mark on the body, he also insisted that, unlike what "those philosophers claim," the body complemented the soul and would exist in heaven.[26] The Platonic ideology he dismissed (as well as his own Manichean training) emphasized the body's inherent evil which only encumbered the soul.[27] For Augustine only corruptibility burdened the soul, which would be remedied in the resurrection. God would restore the body to its ultimate health in Paradise, surpassing even that of Adam and Eve's original form.

Paradise existed—not as an allegory but as a sacred place—even though it also functioned as a didactic tool for spiritual truths; for example, Paradise *signified* the Church and the four rivers, the Gospels. Heavenly fruits and sweet scents *signified* the saints' virtues. That did not mean, however, that Paradise only "signified."[28] Indeed, God would resurrect the flesh and restore it to perfection without even the need to eat. Although Adam and Eve existed in perfect harmony with God, their animal bodies required sustenance and felt hunger and thirst. This spiritual body, transformed and incorruptible, would require no food or drink (even though they might choose to partake). Most importantly, the resurrected form would experience no disparate will, no lust beyond the mind of God. As Paul explained in his letter to the Corinthians (15:42–45):

> So is it with the resurrection of the dead. What is sown is perishable, what is raised is imperishable. It is sown in dishonor, it is raised in glory. It is sown in

> weakness, it is raised in power. It is sown a physical body, it is raised a spiritual body. If there is a physical body, there is also a spiritual body.

Interpreting the apostle's dictum, Augustine taught that the physical body transferred sin and death while the spiritual body conveyed grace and pardon through Jesus' salvific act.[29]

In *City of God* Augustine focuses more on the transformed flesh available in Paradise than its physical pleasures perhaps because he links it to Jesus' own resurrection and salvific act. His theological maxim takes precedence: flesh bares sin's scar—lust—and Christian redemption reverses it. For Augustine, humanity's inability to operate with full reason and will against personal desire and passion defines our "fallen" nature. In Paradise, "God shall rule the man, and the soul shall rule the body."[30] In the original Garden, for example, Adam and Eve would have procreated through sexual intercourse, as purposed in their natural, biological design. Yet sex would have occurred *without* lust or shame; genitalia would have obeyed the will, and Adam would have "sowed seed" in Eve much as farmers plant crops.[31]

Because God transmutes the body into its perfected form within Paradise, righteous citizens of the Heavenly City return to this constant harmony with their bodies and their surroundings. Augustine contends that sexual distinctions will persist; women will be resurrected as females instead of "the image of Christ" (i.e., male). Because no more procreation will occur, sexual intercourse (without lust, of course) becomes redundant. He further describes the new, spiritual body as at the "bloom of youth" or about thirty years old; beautiful and without blemish or fault; capable of eating and drinking yet not particularly enticed by it.[32] He stresses that God can resurrect a perfected body no matter its condition at death; so, those suffering from physical deformities will be "remolded" into their full potential. More importantly, perhaps, those saints who had been devoured by animals in defense of their faith will be gathered together wholly.[33] Not surprisingly then, Augustine largely avoids describing the sensual pleasures that await in Paradise as such joys resonate too much with "earthly" pleasures. Although certainly without pain, sorrow, or struggle, the paradisiacal form reifies the salvific promise instead of enticing with sensory reward.

Ambrose describes Paradise in much more allegorical terms than Augustine. He begins his work *On Paradise* by warning against treating the topic too lightly, perhaps meaning too literally. He notes that not even Paul fully understood his journey recounted in 1 Corinthians 12:1–10 and warned against speaking too freely of the experience. As became common among early Christian authors after Augustine, Ambrose collapsed Paradise, the Garden of Eden, and heavenly Jerusalem. For example, Ambrose interpreted Psalm 46:4—"the stream of the river makes the city of God joyful"—as signifying Eden's paradisiacal waters that fed the eschatological city of Jerusalem.[34] Every aspect of Paradise pointed to a higher spiritual lesson; indeed, Paradise itself represented the soul, fecund in blossoming virtues and the Tree of Life, Wisdom.[35]

Ambrose parallels Eden's four rivers with the four principle virtues; according to Genesis 2:10,

> A river watering the garden flowed from Eden; from there it was separated into four headwaters. The name of the first is the Pishon; it winds through the entire land of Havilah, where there is gold. The gold of that land is good; aromatic resin and onyx are also there. The name of the second river is the Gihon; it winds through the entire land of Cush. The name of the third river is the Tigris; it runs along the east side of Asshur. And the fourth river is the Euphrates.

Ambrose's schema appoints Phison as Prudence; Gihon as Temperance (primarily chastity); Tigris with Fortitude; and, the Euphrates as Justice (the most perfect). He also locates the Garden/Rivers "in the east," signifying the Rising Sun, Jesus.[36] Ambrose emphasizes Paradise's didactic function instead of locating it geographically or defining the sensual pleasures that await the saints.

Some of Paradise's most important lessons relate to sex and gender. According to Ambrose, Adam typified the mind (or, *nous*) while Eve represented the senses and emotion.[37] Eve sinned more egregiously than Adam as she fell prey to Satan, yet she still warranted God's redemption after confession. Ultimately, God anticipated their disobedience because nothing occurs outside His will; and, He *willed* that humanity eventually distinguish between good and evil. In this way, Ambrose includes Eve as a key player in the Eden drama even while recognizing her as "weaker" and "inferior."[38] Eve's most important role, however, rests in her punishment: God ordains that woman should be redeemed through childbearing (Genesis 3:16) and obedience to her husband. This provides for the future generation of Christ in Mary who reverses the original sin and characterizes the submissive relationship between Jesus and His Church.[39]

Eve's role in the Garden of Eden not only shares an emerging Christian theological orthodoxy but also nascent ascetic ideals for females. Early Church Fathers urged pious women to reverse Eve's sin by adopting a chaste life; indeed, Jerome explained that once only great male prophets remained abstinent, but now (in his fifth century) women could maintain such a lifestyle because Eve's curse had been lifted. After Mary birthed Jesus, who reversed the Fall, God bestowed the gift of virginity "most richly upon women."[40] Thus through ascetic practice, women could return to their prelapsarian form.

The ascetic ideals outlined by Jerome focused on transforming the sensual flesh to a new, spiritual self originally present in Paradise. According to

Jerome's *Letter to Eustochium*, humanity's initial fall resulted from appetite: "obeying his belly and not God, [the first man] was cast down from Paradise into this vale of tears." The return to Paradise required avoiding "wine and dainty foods" as well as sexual desires: "abstinence may bring back to paradise those whom satiety once drove out."[41] To achieve such an ascetic ideal, Jerome encouraged his virgins to become like Psalms' "garden enclosed . . . a fountain sealed." They should remain home, dress modestly, and be careful of pride.

An ascetic lifestyle possessed many rewards, however; Jerome's descriptions reinstate a sensual and erotic discourse focused on spirit instead of flesh. Recalling the Song of Solomon, Jerome explains that the waiting virgin's Bridegroom/Jesus shall "put His hand through the hole of the door and your heart shall be moved for him." There, she shall exclaim that "a bundle of myrrh is my well beloved unto me; he shall lie all night betwixt my breasts." Yet Jerome warns Eustochium that "Jesus is jealous," thus she should close her "doors" so that no other man should gaze upon her beauty.[42] Jerome successfully transmutes the virginal body into an enclosed garden, where fecund women produced Godly virtues just as Mary's own womb birthed Christ.

Early medieval Bishop Avitus of Vienne's poetic commentary on Genesis includes many of these same themes. Avitus envisioned Paradise as a place on earth just "beyond where the world begins," yet sealed against humanity because of original sin.[43] Both Adam and Eve once lived in this perfect garden where no winter exists amid an eternal spring; where flowers burst from the earth in vivid color; and cinnamon and balsam sweeten the air.[44] They married with angels' song in the background and Paradise itself as their bridal chamber; yet, they remained innocent and pure, free from lust, greed, or desire.[45] Avitus imagines the return to that state, the life like the angels:

> For . . . there will be no desire for marriage, nor will the joining of flesh bring their passionate sexes together in a disgusting union. Moans will cease and with them debauchery, fear, anger, passion, deceit, grief and treachery, along with sadness, quarrelling and spite. No one will be poor, no one greedy, but under a single peace Christ, the glory of the saints, will answer all our needs.[46]

In Avitus' idealized Paradise, desire disappears along with "negative" deeds and emotions that afflicted the fallen form.

Avitus also envisions the Fall in quite sensual terms: Satan tempts with fruit both delicious and fragrant; Eve tastes the forbidden fruit's sweet venom; and after Adam follows her directives, their bodies felt "indecent impulses."[47] Avitus isolates Eve's weakness and gullibility as the primary factors in the Fall; he also details her punishments as submitting to her husband's desires and childbirth traumas. In his letter to his sister Fuscina, Avitus praises her abstinence as virtue:

> You have chosen to spurn the torches of marriage but to glow with a holy love, to be sluggish in passion but afire in your heart for work. You have chosen to be ignorant of man and to produce the kind of offspring that no sad misfortune can ever take away. You will not weep when deprived of the pledges of your fecund life, nor fear to survive as widow your constant spouse, yourself free from evil. Nor are you touched by the emotion that overcame Eve . . . or a dead child and with it a guilt that lives on afterwards. She was subject to a man and doomed to suffer a master in her chamber. She served in a disgusting bed, as she endured wedlock.[48]

Avitus praises his sister for following Mary instead of Eve; for winning the twin crowns of Virgin and Mother instead of languishing in the death sentence inherited by all.

Ascetic discourse continued to highlight both Paradise and the idealized, prelapsarian body throughout the early Middle Ages. As we have already seen, Christian saints (both living and dead) represented a convergence between heaven and earth, their bodies (and relics) redolent with sweet smells. Those figures displayed the transformed, incorruptible "new" bodies that await Christians in Paradise that, according to Augustine, not only restored but also improved upon Adam and Eve's initial form. Saints' relics thus reminded early Christians of what humanity had lost as well as what would be restored in Heaven; one garden's sin (Adam) restored by another garden's sacrifice (Christ).

Monasteries also prompted a nostalgic gaze back to Paradise. As the ascetic lifestyle organized around various rules and orders, monastics dedicated themselves to a spiritual existence on earth. Even though little Merovingian monastic architecture remains, early Carolingian cloisters already included interior gardens (garths) that effectively recreated Eden for its inhabitants. Generally monks utilized the cultivated space for individual prayer and meditation instead of liturgical rites; however, Sunday processions circumambulated the garden in symbolic recognition of the *axis mundi*, or world center.[49] Cloister gardens became increasingly significant—and ornate—throughout the medieval period.

Christian laity also had access to Paradise within the Church. Incense purified both the altar and the audience as dazzling thuribles hanged from ceilings. Light behind the altar recalled Jesus, Light of the World, and the radiant metals and gems adorning liturgical space harkened to the divine Light.[50] Chant and the spoken Word resonated throughout sacred space; and, the Eucharist tasted sweet to the redeemed.[51] Christian worship appealed to the entire sensorium, providing only a glimpse of the pleasures promised in

Paradise. Even though the earthly body bore sin's mark in its lust and shame, a renovated existence awaited after resurrection.

BODIES PURIFIED: PARADISE AS A PLEASURE GARDEN

Images of Paradise (*al-janna*, or garden) and Hell (*al-nar*, or fire) appear prominently throughout early Meccan sura, hadith, *sira* literature, and eschatological manuals.[52] Islamic scholars generally agree that pre-Islamic Arabian cosmology lacked any coherent notion of afterlife, focusing instead on tribal kinship responsibilities in this current realm. According to the Qur'an 16.38: "And they [people of the *jahiliyya*] swear by Allah their strongest oaths [that] Allah will not resurrect one who dies. But yes—[it is] a true promise [binding] upon Him, but most of the people do not know." Muhammad's message not only of one creator God but also a Final Judgment—culminating in eternal punishment or reward—proved quite shocking and controversial to many dissenters. The notion of an afterlife that focused on implausible, individual, pleasurable rewards/punishments diverged from their dominant communal, tribal ethos.

For others, the graphic and explicit descriptions of eschatological recompense undoubtedly held great appeal. Muhammad taught: "When the sun is wrapped up [in darkness] and when the stars fall, dispersing . . . when the pages are made public, and when Hellfire is set ablaze, and when Paradise is brought near, a soul will [then] know what it has brought [with it]." From the earliest texts, Paradise emerges as very material—believers "bring" their individual deeds with them at Judgment Day and Allah rewards them with luxurious pleasures in Heaven or punishes with Hellfire's tortures.

As Islamic eschatology evolved, then, so did an ethical code which promoted virtuous deeds and material self-denial in this life. Focusing on worldly wealth and reward leads to poverty in the next; focusing on spiritual and virtuous deeds here and now, results in an eternal life of material prosperity. Meccan sura 102.1 explains, "Competing for more distracts you until you go into your graves." According to Anas ibn Malik, for example, the Prophet taught that whoever wears silk and gold in this world will not receive it in the Hereafter.[53] For many pious Muslims, then, such outward displays of wealth and prosperity signal spiritual depravity. Paradisiacal pleasures amplify those available on earth for those who devote their lives in this world to God instead of luxuries.

While we are most concerned with paradisiacal images in this chapter, it is important to note that the depictions of Hell are equally graphic and shocking. Like in the *Apocalypse of Peter*, accounts of punishment and torture abound, usually corresponding with the sins committed on earth. Qur'anic *sura* lay the foundation for the flaming pit where sinners suffer profound thirst, hunger, and pain from scorching water. Sinners' bodies rejuvenate themselves in the Hellfire experiencing tortures and torments in endless cycles of agony. Tradition explicates these tortures in vivid imagery: those who judged unfairly become blind; *`ulama'* who led the community astray chew their own tongues and have pus running from their mouths; perjurers and liars suffer ulcers all over their bodies; slanderers endure their tongues extruding from the back of their necks.[54] Drunkards receive scorching water that cut their intestines and snakes and scorpions that flay their feet. According to some traditions, however, wine drinkers may cry out to Muhammad for mercy and by his intercession be freed.[55] Other hadith suggest that Muslims generally (not just drunkards) will suffer for their sins in the Fire *only temporarily* and then be rewarded with Paradise through Muhammad's intercession.[56]

Islamic traditions regard the Garden of Eden and Paradise with equally vibrant imagery; and, even though they share several themes with the Judeo-Christian narrative, they have various points of departure as well. According to the Qur'an, Allah created Adam and Eve in Paradise and then sent them to the earthly Garden (never named Eden) only after they sinned:[57]

> Then they both ate of that tree, and so their private parts appeared to them, and they began to stick on themselves the leaves from Paradise for their covering. Thus did Adam disobey his Lord, so he went astray. Then his Lord chose him, and turned to him with forgiveness and gave him guidance. Get you down (upon the earth), all of you together, from Paradise, some of you are an enemy to some others. Then, if there comes to you guidance from Me, whoever follows My Guidance shall neither go astray, nor fall into distress and misery. (20.121–23)

Yet as with Christian tradition, the Garden remains difficult to locate geographically. Some theologians imagined that the primordial Garden exists somewhere in Syria, Persia, or India; most often, Islamic tradition conflates the place of exile with the eschatological heavenly realm.[58] Several Arabic terms signify "Paradise" including *al-janna* (garden); *sama'* (heaven); and *firdaws* (paradise).

Notably, the Qur'an never identifies Eve/Hawwa' as primarily responsible for disobeying God's command: "they both ate of that tree" and are thus equally accountable. Hadith and Islamic tradition later emphasized Eve's culpability and the punishment her daughters inherited. In many ways these punishments counter the basic Quranic premise that every human being will be judged equally, reflecting their pious deeds, on Judgment Day. According to al-Tabari, for example, Iblis/Satan first attempted to trick Adam into eating the forbidden fruit, but he failed. He therefore went to Eve and explained the benefits of knowledge and status the tree offered. Eve, the weaker of the two, succumbed to Iblis' temptation and then coerced Adam into eating as well.[59] In recompense for her initial offense, God said:

> Now it is My obligation to make her bleed once every month, as she made this tree bleed. I also must make her stupid, although I created her intelligent (*halimah*), and must make her suffer pregnancy and birth with difficulty, although I made it easy for her to be pregnant and give birth. . . . Were it not for the affliction that affected Eve, the women of this world would not menstruate, and they would be intelligent and, when pregnant, give birth easily.[60]

While Adam and Eve do not bear "original sin" in the same way that Christian theology defines, Adam and Eve's bodies still bear marks—such as sickness, death, and menstruation—that their descendants inherit. God's punishment of Eve (and her daughters) particularly pains the body with menstruation.

Other hadith link God's punishment of Eve's body directly to women's lack of religious piety—and, as a consequence (according to many Islamic scholars), inability to fill roles of public authority. According to Abu Sa`id, the Prophet once taught a group of women:

> "I have not seen anyone more deficient in intelligence and religion than you. A cautious sensible man could be led astray by some of you." The women asked, "O Allah's Messenger! What is deficient in our intelligence and religion?" He said, "Is not the evidence of two women equal to the witness of one man?" They replied in the affirmative. He said, "This is the deficiency in her intelligence. Isn't it true that a woman can neither pray nor fast during her menses?" The women replied in the affirmative. He said, "This is the deficiency in her religion."[61]

This hadith links the ability for women to perform religious requirements not to their faith but to their bodies' impurity.

Several other hadith transmissions accentuate Eve/women's punishments and frailties in comparison with Adam/men. According to Ibn Abbas, Muhammad once had a vision of Paradise and Hell while praying during a solar eclipse. During the vision, he reached his hand out and then withdrew it. After the trance-like state ends, his followers asked what he was doing; he responded that he reached for a bunch of giant, paradisiacal grapes. He then described his vision of Paradise and the Fire. He related that Paradise consisted mostly of the earthly poor, and women comprised the majority of hell's population. When asked to explain, he blamed women's "ungratefulness" to their husbands.[62]

Later *mi`raj* narratives described not only Muhammad's visit to Paradise but also his tour of Hell—which occurred either in a dream or in his body—in vivid detail. The *mi`raj* account, woven together from various strands, outlines the Prophet's journey to meet God thus providing a heavenly topography which largely shapes the Islamic view of Paradise. For example, according to early accounts such as those by Ibn Ishaq in the *sira*, Muhammad travels through seven heavens passing other monotheistic prophets who recognize his superiority on the way. In the lowest heaven, his guide Gabriel introduces him to Malik, the keeper of hell. Then, the Prophet views the torments of hell and its assorted group of sins including stealing from orphans, usury, and adultery. Over time, however, an emphasis on women's sins evolves stressing particularly sexuality and speech. These traditions correlate women's eternal fate with her submission not only to God but also her husband and community (e.g., appearing uncovered, singing, and going outside without permission).[63]

Other hadith link women's presence in Paradise to their role as mother: "the woman who has just given birth, her child shall drag her on the Day of Resurrection by her navel-cord into the Garden."[64] While Islamic texts carefully illustrate men's paradisiacal sexual pleasures, the occasional mention of women's roles in Heaven usually relate to children, family, and childcare.[65] Indeed, while men enjoy a variety of sexual partners, women gain contentment (and sexual satisfaction) from their one husband which again mirrors the Arabian polygamous family ethic.[66] Even though the Qur'an stipulates that God will judge men and women equally, Islamic tradition perhaps more aptly reflects its patriarchal cultural milieu and imagination.[67]

Early Islamic authors go on to describe Adam and Eve's life after their expulsion from Paradise. While Adam might be remembered most for disobeying God, Muslims also celebrate his repentance and submission to Allah (as a muslim, or "submitter" to God). The *Tales of the Prophets* commemorate Adam and Eve's loss of God's Garden, rich in sensual beauty, because of their defiance and Iblis' (or Satan's) trickery. Iblis tempted Eve by appealing to her senses: "Look at this tree! How sweet does it smell! How good does it taste! How beautiful is its color!"[68] The Garden contained perfect life and beauty—without stain, impurity, death, or decay—and consisted of gold and silver brick, musk mortar, and saffron soil. One of the things Adam mourned losing the most after his expulsion was the sweet "smell of the Garden and its perfume."

Because of his attachment to paradisiacal scent, the *Tales* agree that Adam originally brought perfume to earth, although they offer variant descriptions as to how. One unifying theme is that after falling from Paradise, Adam landed in India and Eve in the Arabian Hijaz. According to Abu l-`Aliyah, Adam took a branch from the Garden and made a wreath for his head; when he came to earth, the leaves wilted and scattered, resulting in all aromatic plants located in India that provide the source for perfume. Another tradition traced through Ibn Abbas explains that Adam brought various aromatic trees from Paradise and planted them in India. A final transmission through Sufyan, details Adam's elaborate garment made of leaves that dried and became the source of aloe trees, sandalwood, musk, ambergris, and camphor trees. Al-Tha`alabi admits that one perceptive audience member asked Muhammad whether musk and ambergris originated from animals instead of aromatic plants (which they do). The Prophet skillfully responded that, indeed, musk originated from one "resembling a gazelle," but the animal feeds on aromatic bushes and God then transforms the plant life to musk in its navel.[69] Likewise, the "sea animal" that produces ambergris once grazed on land in India and then God (via Gabriel) drove her into the sea.[70] Muhammad's explanation allows for musk and ambergris' paradisiacal origin as plants, later transported to earth by Adam.

Al-Kisa'i connects Adam and Eve with perfume in even more poetic traditions. According to his *Tales*, perfume (specifically musk) sprang from Adam's tears as he wept for all he had lost in Paradise. He also describes the locales of pre-Islamic Mecca, which had been blessed by God above all other earthly sites. The Zamzam well in Mecca, for example, held water that smelled sweeter than musk; and, as Eve bathed in the well, the scent wafted throughout the earth.[71]

Adam and Eve's exile from Paradise did not detract from its beauty which, according to Islam, still awaits pious believers. Unlike much of their Judeo-Christian and even Zoroastrian neighbors, early Islam promoted a much more opulent materialism in its paradisiacal traditions. The Qur'an describes a lush Garden with rivers of wine, milk, and honey (47.15); couches where the righteous recline, eating and drinking their fill; air redolent with sweet scents such as camphor, musk, ginger, and saffron; and structures built with precious metals and gems. Recreating its Arabian social context that allowed for slavery, Allah also provides scores of male servants to attend to the believers' needs. These servants appear to be a human species created solely for reward, distinctive from humans who function with free will (however limited). Another uniquely paradisiacal species—heavenly *hur* (black-eyed virgins)—provide pious men with limitless sexual fulfillment. Hadith and later eschatological manuals build upon this Qur'anic base, exploring and explaining the details of paradisiacal bliss. Descriptions of houris, for example, appear as rewards for Muslim martyrs who fight against their earliest enemies in Mecca.[72] Muhammad also encourages his supporters to battle with "hope of reward in heaven."[73]

Various hadith notably focus on sex and sexuality. According to many traditions, God provides increased sexual prowess and immense sexual pleasure for men with their wives, houris, and (perhaps) young men.[74] God creates houris—a unique heavenly species—from light precisely for men's pleasure. While the houris bodies certainly bestow sexual satisfaction, their pious male mates enjoy simply gazing upon their beauty; admiring their spangled and bejeweled limbs; and, inhaling their sweet scent. These females live only for their husbands (pious Muslim men) and remain virginal even after intercourse.[75] The houris engage all the senses especially sight and scent: men gaze upon the females' white yet translucent skin and see their own faces reflected.[76] They have their noble man's name written on their breasts and wear precious gems and jewelry on every limb.[77] God also created their bodies of scent: from toes to knees is saffron; knees to breasts is musk; breasts to necks of amber; and necks to heads is camphor.[78]

Islamic tradition fully explores the events of Judgment Day and who will be present so that a fundamental (although not uncontested) script of the afterlife exists by the end of the twelfth century. Even though the traditions vary in detail, a general order appears.[79] At death, the angels Munkir and Nakir document the soul's deeds, both good and evil. Angels also question the soul and then lead it to Hell or Heaven; they generally pass through seven gates on their way to Paradise. Some souls remain asleep in the tomb or wander about the earth.[80] Then at the end times, God resurrects all beings that have ever lived (and some animals that have been sacrificed) and they proceed before Him and His Judgment Scales. Each human being stands before God with her deeds weighed; if the scales tip with good deeds, she inherits Paradise, bad deeds doom to Hell. Muhammad leads his community over the Sirat Bridge into the Garden; according to some traditions, those who lacked faith or failed to follow the *shariah* may fall from the bridge into Hell. In contrast, Muslims who faithfully gathered at mosques may ride into the Garden on beautiful camels, necks like saffron and heads fragrant with musk. They find there the River of Mercy with pearl pebbles, mud that smells of musk, and saffron straw.[81] Many *`ulama'* who lived piously receive seats of light; martyrs and "virtuous" sit on mounds of musk.[82] Clouds rain rosewater upon camphor hills and saffron plants.[83]

As transmitted by Jabir ibn `Abdullah, Ridwan (the angel tasked with maintaining Paradise) takes special care of fasters who, from their graves arrive hungry and thirsty. He greets them with grilled meat and fruit and presents them with castles composed of pearl, ruby, chrysolite, gold, and silver.[84] Other groups singled out for reward include prophets, warriors, scholars, martyrs, those who know the Qur'an, Imams, muezzine (those who give the call to prayer), women who die in childbirth, those killed unjustly, those who died on Friday; those who were patient and those who loved God.[85] As the righteous enter the Garden, they drink from its fountains, making their "bodies pleasant like musk." Their bodies no longer urinate or defecate; instead, after they eat and drink, "it comes out from their bodies like the scent of musk." They neither spit nor blow their noses; they do not ejaculate in sexual intercourse even though they experience immense pleasure.[86] Men's sexual potency increases as each one equals one hundred men of the world, capable of sexual intercourse with one hundred virgins daily.[87] Wives no longer menstruate.[88] The Garden provides pleasures beyond imagination, yet without any associated corporal pollutions.

Islamic Paradise encompasses a feast of physical pleasures; amusements enjoyed by the righteous on earth, both sensual and sexual, impart only a shadow of what awaits in Heaven.[89] Many anti-Islamic authors and Orientalists have historically critiqued this very sumptuous display of Paradise. Some Muslim authors even question the validity of such images and consider them fit only for "uncouth peasants," or suggest they should be considered strictly as allegory.[90] While that approach would be disingenuous to the majority of early Islamic textual traditions, it is important to consider this lavish view of heavenly reward with two points in mind.

First, Paradise's sensorial feast should be considered in its fullness; instead many Western authors focus disproportionately on sex and sexuality (and especially the houris). Taken as a whole, paradisiacal descriptions emphasize sensory fulfillment—the body experiences every sense, every source of pleasure, perpetually in heaven. Food once taken from a bowl instantly replenishes itself; water and wine (that does not intoxicate) flow without end. This rhetoric of pleasure highlights the eternal nature of the afterlife as well as the body's completion and perfection; time has no real meaning. It also perfectly complements hell's perils and torments, clarifying each realm by their juxtaposition.[91] Altogether paradisiacal luxuries express God's Glory because, according to tradition, all sensory indulgence pales in comparison to gazing upon the Beatific Face.[92]

Al-Ghazali explores his specific ranking of pleasures in his *Book of Love, Longing Intimacy, and Contentment*. He writes that God created each faculty for its complementary pleasure: thus, man takes pleasure in eating because it satiates his hunger; the vengeful take pleasure in triumph because it satiates his anger. All humanity's faculties link with the senses, resulting in both pleasure and pain. For al-Ghazali, the intellect (*`aql*) is the most important human faculty; it resides in the "heart," and comprehends things beyond simple sensory knowledge. Al-Ghazali also ranks the "pleasures of knowledge"; for example, understanding farming is excelled by knowing how to rule or order a kingdom. Wise men and women should cultivate the inner faculties of "knowing" over the outer pleasures of sensory indulgence; and, ultimately the inner knowledge of God over every other knowledge. Once the seeker truly knows God—tastes Him (*dhaqa*)—she experiences true bliss.[93] Paradise allows the pure of heart to feast upon God's presence.

For al-Ghazali, then, sensual paradisiacal pleasures are the "lesser" order, rewarding the pious for their obedience. But for those who loved God not because they desired reward or feared punishment, gazing upon God with a whole heart (combining both the inner and outer ways of knowing) makes everything else pale in comparison. According to al-Ghazali, "Were all the pleasures of this world to be spread out for them at that moment [of encountering God], they would not spare them a glance because they possess consummate bliss and that utmost joy which is limitless."[94] In Paradise, some will take joy in the pleasures offered in Paradise, such as eating, drinking, and sexual intercourse; others, those who perfected their "knowing" while still on earth, find absolute pleasure in gazing upon God. Paradise as sacred space and time remains just beyond imagination; trees' shadows stretch for miles and rapturous pleasures last for thousands of year without participants even noticing. The imaginable diminishes alongside the ineffable presence of God.

Ghazali's insistence upon the paradisiacal gaze upon God's Beatific Face is extremely important and controversial. Islamic theologians argued over God's absolute transcendence. Many taught against any form of anthropomorphism which occurred in describing one's love for the Divine. This discursive mode allowed for objectifying—and humanizing—God and, perhaps, admitting His lack (e.g., his loneliness, or need for admiration). Thus for theologians like Abu al-Qasim al-Zamakhshari (d. 538/1144), contemplating a "love" relationship between the self and God was blasphemous; the appropriate relationship was between master and servant, creator and created. But for al-Ghazali, pious men and women love God by "tasting" Him—knowing Him fully with their heart—and finally encounter Him in paradise, when He lifts the veil and perfects and purifies the soul.[95]

This leads us to the second important point in understanding Paradise in Islam: what role does the body play in Islam as a whole? Paradisiacal existence certainly amplifies (or perfects) the body's enjoyments of sensory pleasures; paradisiacal bodies even appear at their healthiest age, usually thirty to thirty-three.[96] After eating and drinking their fill, "their bellies become slender once more" (with no need to worry about weight gain).[97] This contrasts significantly with the transformed, spiritual body presented in Christian heavenly narratives. The body in Islam does not require redemption from original sin; it demands perfection from immortality and impurity. God does not punish the body with sensual desire and physical pleasure, which the Christian heaven transforms. God gifts the body with these delights and then perfects them in Paradise.

While Islam and Christianity understand the body differently, both traditions rely upon scent to depict paradisiacal favors reserved for pious men and women. Paradise's landscape offers abundant sweet smells ultimately associating this archetypal Garden with its earthly echoes. Roman courtyards, monastic enclosures, and Islamic pleasure gardens all provide but a glimpse of eternal sensual reward. Believers' transformed or perfected bodies demonstrate that death, decay, and mortality have no place there. Fragrance, in its unique way, thus extends beyond paradisiacal boundaries into earthly time and space, reminding humanity not only of perfection lost but also perfection promised.

NOTES

1. See the fine article by Martin Goodman, "Paradise, Gardens and the Afterlife in the First Century CE," in *Paradise in Antiquity: Jewish and Christian Views*, eds. Markus Bockmuehl and Guy G. Strouma (Cambridge, UK: Cambridge University Press, 2010), 57–63.

2. *1 Enoch* 24–27. Kelley Coblentz Bautch provides an important study of Enoch's geography in "Situating the Afterlife," ed. April D. DeConick, *Paradise Now: Essays on Early Jewish and Christian Mysticism* (Atlanta: Society of Biblical Literature, 2006), 249–64.

3. *1 Enoch*, 29–32.

4. *4 Ezra* 2.12–13 also associates tree of life with perfume and immortality. God warns Ezra that the "others" will receive the "everlasting habitations" that He originally prepared for Israel. He will do so by giving the "others" the tree of life and its perfume. Scholars disagree about the occurrence of two "Gardens" in *1 Enoch*. One argument suggests an attempt to reconcile the Biblical appointment of Paradise in the East, on a mountain, and in the North (e.g., Genesis, Isaiah 14, and Ezekiel 28). Another argument suggests independent traditions that emphasize God's provision of knowledge and insight to the "righteous" (and Adam) instead of concern for paradisiacal geography. See Eibert J. C. Tigchelaar, "Eden and Paradise: The Garden Motif in Some Early Jewish Texts (I Enoch and Other Texts Found at Qumran)," *Paradise Interpreted*, 37–62.

5. *Life of Adam and Eve*, in *The Old Testament Pseudepigrapha*, ed. James H. Charlesworth (New York: Doubleday, 1983–1985), vi–x.

6. *Life of Adam and Eve*, xxiv.

7. *Life of Adam and Eve*, xxix, 1–7.

8. *Life of Adam and Eve*, lxxviii–xli.

9. *2 Enoch* 31.1, 42; 8.1–3.

10. *2 Enoch* 8.5–6.

11. *2 Enoch* 22.

12. *3 Enoch*, 48A.1. Also, see Ira Chernus' discussion of human's ability to gaze upon God in "Visions of God in Merkabah Mysticism," *Journal for the Study of Judaism* XIII.1–2 (1982): 123–46.

13. *3 Enoch* 1.6–8; 48A.2.

14. James Tabor situates Paul's ascent text in historical and cultural context in *Things Unutterable: Paul's Ascent to Paradise in Its Greco-Roman, Judaic, and Early Christian Contexts* (Lanham: University Press of America, 1986).

15. *The Apocalypse of Paul*, 14–18.

16. *The Apocalypse of Paul*, 19.

17. *The Apocalypse of Paul*, 22–28.

18. *Apocalypse of Paul*, 31–39.

19. *Apocalypse of Peter*, 7, 8.

20. *Apocalypse of Peter*, 16.

21. Augustine, *Confessions*, 8.12.29.

22. Augustine, *Confessions*, 8.12.29.

23. Augustine, *Confessions*, 8.12.30.

24. Markus Bockmuehl discusses how early and medieval Christians combined the celestial and terrestrial "ideas" of paradise in his important essay, "Locating Paradise," in eds. Markus Bockmuehl and Guy G. Stroumsa, *Paradise in Antiquity* (New York: Cambridge University Press, 2010), 193–209.

25. Augustine, *City of God*, XIII.13.

26. Augustine, *City of God*, XIII.17–20.

27. See Robert Hauck's important article about early Christian notions of the body (and its senses) in dialogue with their Platonist counterparts. He touches on Augustine's teachings even though he focuses primarily on Origen and the author of Clementine's Homily 17. "They Saw What They Said They Saw: Sense Knowledge in Early Christian Polemic," *Harvard Theological Review* 81:3 (1988): 239–49.

28. Augustine, *City of God*, XIII.21.

29. Augustine, *City of God*, XIII.22–23.

30. Augustine, *City of God*, XIX.27.
31. Augustine, *City of God*, XIV.23–24.
32. Augustine, *City of God*, XXII.12–15, 19.
33. Augustine, *City of God*, XXII.16, 21.
34. Ambrose, *On Paradise*, 1.4.
35. Ambrose, *On Paradise*, 3.13; 1.6.
36. Ambrose, *On Paradise*, 3.25–28.
37. Ambrose, *On Paradise*, 2.8–9.
38. Ambrose, *On Paradise*, 4.24.
39. Ambrose, *On Paradise*, 10.47.
40. Jerome, *Letter XXII*, 21.
41. Jerome, *Letter XXII*, 10.
42. Jerome, *Letter XXII*, 25.
43. Avitus, Poem 1.188–232; trans. 77.
44. Avitus, Poem 1.233–81; trans. 78.
45. Avitus, Poem 1.188–232; trans. 77.
46. Avitus, Poem 2.1–42; trans. 80.
47. Avitus, Poem 2.260–303; trans. 86.
48. Avitus, Poem 6.130–170; trans. 137.
49. See Mary Helms' important article, "Sacred Landscape and the Early Medieval European Cloister: Unity, Paradise, and the Cosmic Mountain," *Anthropos* 97.2 (2002): 435–53.
50. See Mary Helms' important article, "Ineffable Illumination: Early Medieval Church Treasure and the Preservation of Heaven's Light," *Anthropos* 109 (2014): 103–18. Herein, she explores the meanings of precious metals and gems as not only displays of political and ecclesiastical authority but also symbols of life and cosmic sustenance (light).
51. In her work "Exploring the Sense-Scape of the Gospel of Mark," Louise Lawrence focuses on sense and text (instead of liturgical process). However, she notes that Mark's rebellious focus on hearing/speech diminishes the Imperial prominence of gaze/sight. The Christian "good news" was *heard*, not just displayed. *Journal for the Study of the New Testament* 33.4 (2011): 387–97.
52. Rustomji provides an excellent survey of early eschatological manuals in *The Garden and the Fire*, chapter 6.
53. See Bukharī, 77.48. Other prohibited worldly luxuries included brocaded cloth, gold and silver. As Rustomji points out, this early Islamic ethic might betray an anti-Persian bias as Sassanian Iran was known for its luxury items. See *The Garden and the Fire*, 54–55.
54. Al-Qadi, *Islamic Book of the Dead*, 77–9.
55. Al-Qadi, *Islamic Book of the Dead*, 117–18.
56. Al-Qadi, *Islamic Book of the Dead*, 122–23. See Binyamin Abrahamov's article, "The Creation and Duration of Paradise and Hell in Islamic Theology," *Der Islam*, 79 (2002): 87–102. Abrahamov reviews a variety of traditions that describe Heaven and Hell as temporary abodes; others that claim that unbelievers will remain in Hell forever while Muslims will be released after their appropriate punishments. These beliefs appear particularly among Mu`tazilites and the Khārijites.
57. Tradition relates that God created Adam and cast him out of Paradise on Friday, the "lord" of the week. The resurrection (the Hour) will also occur on a Friday. See al-Ṭabarī, *History*, 112, p. 283.
58. Sri Lanka boasts "Adam's Peak," a pilgrimage site for many Muslims today. It is also important for Buddhists who believe the site holds the Buddha's footprint. Abrahamov also reviews disparate traditions that place the Garden in heaven/earth; and understand it to be a past/future creation; pp. 87–91.
59. According to one tradition, Eve gave Adam wine to drink, thus impairing his reason; only when he became drunken did he eat from the tree. See al-Ṭabarī, *Ta'rīkh al-rasul wa'l-mulūk*.
60. Al-Ṭabarī, *Ta'rīkh al-rasul wa'l-mulūk*, 109; pp. 280–81.
61. *Bukharī* 6.9.
62. *Bukharī* 2.22. Most of the major hadith collections contain a version of this story.

63. See Robert Tottoli's "Tours of Hell and Punishments of Sinners in *Mi`raj* Narratives: Use and Meaning of Eschatology in Muhammad's Ascension," *The Prophet's Ascension: Cross-Cultural Encounters with the Islamic Mi`raj Tales* (Bloomington: Indiana University Press, 2010), 11–26. Tottoli suggests that women's sins were emphasized, partially, as critique of their public rituals celebrating the Prophet's ascension. Also, it is important to note that Muhammad warns against unlawful speech and sexuality for both genders; however, these narratives' attention to women's action appears out of proportion to that of men. For example, *Muwatta' Malik*, 56.11: "Whomever Allah protects from the evil of two things will enter the Garden. They are what is between his jaws and what is between his legs."

64. Al-Tayālisī, *Musnad*, no. 578; quoted in Smith and Haddad, "Women in the Afterlife: The Islamic View as Seen from Qur'an and Tradition," *Journal of the American Academy of Religion* 43.1 (1975): 39–50.

65. See Smith and Haddad's discussion, "Women in the Afterlife," 46–48.

66. Many hadith address the issue of women who have multiple husbands because of divorce or widowhood. They generally agree that the woman's paradisiacal mate would be the first husband. See related discussion in Rustomji, *The Garden and the Fire*, 111–12; Smith and Haddad, "Women in the Afterlife," 46.

67. See, for example, the principle of equality in Qur'an 4.124: "And whoever does works of righteousness, male or female, and is a believer, will enter the Garden and will not be wronged in the least."

68. Al-Ṭabarī, *Ta'rīkh al-rasul wa'l-mulūk*, 105–6; pp. 276–77.

69. See "Cain and Abel" in Aḥmad ibn Muḥammad Al-Tha`alabī, *Arais Al-Majalis Fi Qisas Al-Anbiya*, trans. William M. Brinner, *Lives of the Prophets* (London: Brill, 2002), 6, p. 61.

70. Al-Tha`alabī, "Cain and Abel," 6, p. 61.

71. Al-Kisa'i, 29, p. 67.

72. See, for example, *Sīrat* 546, p. 349–50, and the Battle of Badr. Rustomji discusses numerous such passages in *The Garden and the Fire*, 14–19.

73. For example, *Sīrat* 696, p. 450–51.

74. For example, al-Ghazālī refers to heavenly beings "*ḥūr*" and "*wuldān*" (beautiful young men) in *Durra* 40, p. 45; he also notes that Uriah receives *ḥūr* and beautiful boys as compensation for David's sins against him; 75, p. 66–67.

75. Al-Qadi, *Islamic Book of the Dead*, 124.

76. Al-Ṭabarī, *Tafsīr* v. 23, 24.

77. Al-Qadi, *Islamic Book of the Dead*, 130–32.

78. Al-Sha`rānī, *Mukhṭaṣar al-tadhkira al-qurṭubīyya*, Cairo, n.d., 103; al-Qadi, *Islamic Book of the Dead*, 130–31.

79. For an important overview of the apocalyptic narratives' complexities and historical contexts, see David Cook's *Studies in Muslim Apocalyptic* (Princeton: Darwin Press, 2002).

80. Al-Ghazālī, *al-Durra* 35–37; pp. 43–53.

81. Al-Qadi, *Islamic Book of the Dead*, 101–3.

82. Al-Ghazālī, *Kitab durra*, 107; p. 87.

83. Al-Ghazālī, *Kitab dhikr*, p. 233.

84. Al-Qadi, *Islamic Book of the Dead*, 83–84.

85. Al-Qadi, *Islamic Book of the Dead*, 84, 88.

86. Al-Qadi, *Islamic Book of the Dead*, 133.

87. Al-Ṭabarī, *Tafsīr*, v. 29, 137.

88. Al-Ghazālī, *Dhikr*, p. 245.

89. Al-Ghazālī discusses the promise full gratification in Paradise in *Kitab al-Maḥabba wa'l-uns wa-l-shawq wa'r-riḍa*, Cairo, 1961, 52. Also discussed in Aziz al-Azmeh's "Rhetoric for the Senses: A Consideration of Muslim Paradise Narratives," *Journal of Arabic Literature* 26.3 (October 1995): 215–31.

90. Consider the works of Ibn al-Rāwandī (d. c. 860) and Miskawayh (d.c. 1030); discussed in al-Azmeh, 219–20.

91. The Fire's torments appear equally perpetual, for example: fire and scalding obliterate flesh that regrows and suffers again; inhabitants climb for years to reach scalding mountains' tops only to fall down and restart. See al-Ghazālī, *Dhikr*, pp. 219–31.

92. See al-Ghazālī, *Kitab muḥabbah,* chapter 5, pp. 56–66, "An Exposition of Why the Beatific Vision in the World to Come Surpasses Knowledge in this World."
93. al-Ghazālī, *Kitab muḥabba,* chapter 4, pp. 42–50.
94. al-Ghazālī, *Kitab muḥabba,* chapter 4, p. 53.
95. Al-Ghazālī notes, however, that "direct seeing in the next world" can also be experienced in this world, differing only in "magnification and lucidity." *Kitab muḥabba,* chapter 5, pp. 58–59.
96. Rustomji perfectly employs the term "amplifies" regarding heavenly pleasures.
97. al-Ghazālī, *Kitab dhikr*, p. 243.

Conclusion

Throughout these chapters, I have borrowed methodologies from cultural history, sociology, anthropology, archeology, and gender studies to understand the worlds of early Christianity and Islam. This is not so unusual in Religious Studies which is one reason why I am happily affiliated with it—Religious Studies is messy, especially *Comparative Religion*. Disciplinary boundaries bend, category gives way to diversity, and faith/text/ritual contend for center stage. Comparison particularly allows for new insights into religious belief and practice through identifying similarities as well as differences.

In our globalized world of religious pluralism, observing the past through such a lens provides a useful tool for engaging our contemporary world as well. I certainly do not claim that my arguments above apply only to Christianity and Islam; most world cultures and religious systems associate sweet scent with the transformation of suffering, for example. I also do not claim that my arguments above apply only to late antiquity and the Middle Ages; we need only observe the commodification of fragrance in contemporary America to imagine scent's connection with class and gender.

This work on *Sacred Scents* focuses particularly on embodiment and space—additionally, messy topics. Human bodies, across time and space, experience the world through similar sense organs yet interpret their inputs in very different ways. Traditionally, scholarship has privileged sight/text and reason in cultural studies; what do people see, read, think, believe? While important, certainly, this approach neglects the other senses in ancient and medieval epistemology. Bodies smell the world around them, and those scents *mean*—whether they encode religious identity, class, or gender. Indeed, scent serves as an apt signifier as fragrance itself is messy—crossing boundaries and defying points of origin, scent generally only "smells like" its referent.

Bodies also exist in space, which may be identified as sacred/profane, public/private, and masculine/feminine at different times for distinctive purposes. Material culture certainly marks sacred space; for example, consider placing a crucifix, crescent moon, Bible, or Qur'an within a room. That space transforms; it bears religious meaning. Sacred scent functions in the same way: it transmutes bodies from sinner to saint; it offers healing aroma for both body and soul; it identifies both gender normative and queer. As the ultimate transgressive sensory source, aroma heralds transition and transformation.

In the previous chapters, we have explored what the sensory worlds of early Christianity and Islam *meant.* Experiencing sweet smells—whether prized incense or musk—provided a sensual feast that brought pleasure, increased the libido, healed the soul and body, and excited taste. From Roman and Middle Eastern bath cultures to complex recipes for the kitchen, late antiquity's inclusion of sweet scent—and the stench of "others"—informed how bodies encountered and transmitted social and cultural meaning.

Scent as signifier has a complicated history for early Christians and Muslims. In the Roman world, spices and perfumes were once widely available although quantity and quality distinguished the wealthier classes. As the Roman trade networks devolved, access dwindled: aromatics increasingly signaled episcopal authority as incense wafted through sacred space and royal patrons provided monastic kitchens with spice endowments. Fragrance remained generally accessible in the more urban, Islamic world and its proper use characterized the cultivated Arabs from the barbarous Bedouin. For a time. Yet, just as scent wafts and wanes so too does social and political discourse regarding its meanings. Roman moralists opined the elite's passionate obsession with sensual gratifications; Church Fathers forbade Christian believers to participate in the city's odiferous pleasures; and critics labeled wealthy Muslim dynastic courts as decadent sinners.

Perhaps the most unstable and potentially vulnerable cultural identifier referenced by scent is gender and sexuality. Men in late antiquity and the early Middle Ages perfumed their bodies and their clothes but could be publically shamed and even vilified if they wore the "wrong scent." In both Christian and Islamic patriarchal systems, women's bodies in particular instantiated family and social virtue as well as shame. Male authors celebrated the sweet smell of mothers and wives but lamented women's existential and spiritual threat when they wore "too much scent." Christian exegetes defined Eve's original sin as one of appetite and sensual desire as she smelled the sweet, forbidden fruit and gazed upon its beauty. Both religious traditions, in their distinctive theologies, promote all women as the inheritors of Eve's carnal proclivities.

Early Christians and Muslims also applied their sensual strategies when imagining the Divine; just as they relished the aromas and flavors available through trade networks, they recognized God's own enjoyment of fragrant, virtuous deeds. Modeled after sacrificial offerings, pious lives and sacred performances wafted to Heaven; and, in both traditions, saints functioned as their own trade network, mediating between earth and eternity. Saints' and their dead body parts (at least in Christianity) signal the paradox of both earthly and heavenly habitation—the holy man or woman resides in Paradise yet the body remains in the world. Both time and space collapse with the liminal space of Paradise-imagined.

Paradise regained, of course, informs most Christian and Islamic scent traditions. According to textual descriptions as well as artistic depictions (which we were unable to discuss in this work), Paradise abounds in sweet scent, running waters, and lush foliage. Christian discourse focused on the paradisiacal body transformed, healed of sin, and liberated from the carnal desires of worldly flesh. Salvation and baptism effectively transmuted the corporal senses into spiritual organs, attuned to virtue instead of vice. Muslim Paradise, in comparison, contained believers purified of sin but also defilement. Paradisiacal bodies celebrate with fine drink, good food, and great sex. Physical pleasures, originally gifted from God, exist without concomitant pollutions such as urination, defecation, ejaculation, or other excreta. According to many Muslim thinkers, however, all of these gratifications pale, finally, by the transcendent presence of God. The body communes with the Divine while imparting His ineffable qualities across time and space.

In the end, it is my hope that this work contributes a comparative perspective in its own messy voice to an already vigorous discussion about religion, embodiment, and the sensorium. Focusing on late antique and early medieval Christianity and Islam allows us to follow an emerging discourse relating to the sensorium in two nascent religious communities redolent with social, political, and cultural transition. Indeed, such a conversation seems crucial as contemporary society reimagines sexuality and marriage, women's control over their own bodies, racial violence, and extremist religious identities. Particularly in our postmodern culture, conversations about shared embodied experiences, even interpreted in divergent ways, reminds us of our common humanity.

Works Consulted

PRIMARY SOURCES—EDITIONS AND TRANSLATIONS

Acts of Peter. Trans. J. K. Elliott, *The Apocryphal New Testament*. Oxford: Oxford University Press, 2005.

Acts of Thomas. Trans. J. K. Elliott, *The Apocryphal New Testament*. Oxford: Oxford University Press, 2005.

Apocalypse of Paul. Trans. Bart D. Ehrman. *Lost Scriptures: Books That Did Not Make It into the New Testament*. Oxford: Oxford University Press, 2002.

Apocalypse of Peter. Trans. Bart D. Ehrman. *Lost Scriptures: Books That Did Not Make It into the New Testament*. Oxford: Oxford University Press, 2002.

Ambrose. *De Virginibus. PL 16*. Trans. Rev. H. DeRomestin. *Concerning Virgins. Early Church Fathers: Nicene and Post-Nicene Fathers, Second Series*, v. 10. www.ccel.org.

Anthimus. *De observatione ciborum*. Trans. Mark Grant. *On the Observance of Food*. Totnes: Prospect Books, 1996.

Apicius. *Apicius: A Critical Edition With An Introduction And An English Translation Of The Latin Recipe Text*. Trans. Christopher Grocock and Sally Granger. Totnes, UK: Prospect Books, 2006.

Apostolic Fathers. Trans. Kirsopp Lake. *LOEB Classical Library*. Cambridge, MA: Harvard University Press, 1985.

The Apostolic Tradition. Eds. Harold W. Attridge, Paul F. Bradshaw, Maxwell E. Johnson, and L. Edward Phillips. Minneapolis: Fortress Press, 2002.

Aristotle. *Eudemian Ethics*. Trans. Anthony Kenny. Oxford: Oxford University Press, 2011.

———. *Nicomachean Ethics*. Trans. Terence Irwin. Cambridge, MA: Hackett Publishing Company, 1999.

Athanasius. *Vita Antonii. PG* 26.835–976. Trans. Robert C. Gregg. *The Life of Antony and the Letters to Marcellinus: Classics of Western Spirituality*. Mahwah, NJ: Paulist Press, 1980.

———. *Second Letter to Virgins*. Trans. David Brakke. *Athanasius and the Politics of Asceticism*. Oxford: Clarendon Press, 1995.

Augustine. *City of God. CCSL* 47–8. Ed. Philip Schaff. *Early Church Fathers: Nicene-Post-Nicene Fathers, Series 1*, v. 2. www.ccel.org.

———. *Contra Faustum. CSEL* 25.1. Trans. Rev. Richard Stothert. *Against Faustus. Early Church Fathers: Nicene-Post-Nicene Fathers, Series 1*, v. 4. www.ccel.org.

———. *Confessiones. CCSL* 27. Trans. Henry Chadwick. Oxford: Oxford University, 2009.

———. *Contra Academicos. CSEL* 63.3–81. Trans. John Joseph O'Meara. *Against the Academics. Ancient Christian Writers Series*, v. 12. Mahwah, NJ: Paulist Press, 1951.

———. *De fide et operibus*. Trans. Gregory J. Lombardo. *Ancient Christian Writers*, v. 48. Mahwah, NJ: Paulist Press, 1988.
———. *De paradiso. CSEL 32.1.265–336*. Trans. J. J. Savage. *On Paradise: Fathers of the Church*, 42. Washington, DC: Catholic University of America, 1961.
———. *Enarrationes in Psalmos. CCSL* 38–40. Ed. Philip Schaff. *Expositions on Psalms. Early Church Fathers: Nicene-Post-Nicene Fathers, Series 1*, v. 8. www.ccel.org.
———. *Epistulae. CSEL* 57. Ed. Philip Schaff. *Letters. Early Church Fathers: Nicene-Post-Nicene Fathers, Series 1*, v. 1. www.ccel.org.
———. *Sermones. PL* 38–9. Trans. Mary Sarah Muldowney. *Fathers of the Church*, V. 38. Fredericksburg, PA: Catholic University of America Press, 1959.
———. *In Joannis Evangelium ractatus. CCSL* 36. *Tractate on the First Epistle of John 6:9*. Edited by Peter Schaff. *Early Church Fathers: Nicene-Post-Nicene, Series 1*, v. 7. www.ccel.org.
Avitus. *Poems*. Trans. George W. Shea. Tempe: Arizona State University, 1997.
Al-Baghdadī, Ibn al-Karīm. *Kitāb ṭabīkh*. Trans. Charles Perry. *Cookery Book*. Totnes: Prospect Books, 2005.
Benedict of Nursia. *Regula*. Trans. Luke Dysinger. *Rule of St. Benedict*. Naperville, IL: Source Books, 1997. (Latin and English).
Al-Bustī, Muḥammad ibn Ḥibbān ibn Aḥmad Abū Ḥātim al-Tamīmī. *Kitāb al-majrūḥīn min al-muḥaddithīn wa'l-ḍu'afā'wal'l-matrūkīn*. Ed. Maḥūd Ibrāhim Zāyid. Aleppo, 1975.
Cassian. *De institutis coenobiorium. CSEL* 17. Trans. Boniface Ramsey. *Early Church Fathers: Ancient Christian Writers Series*, v. 58. Mahwah, NJ: Paulist Press, 2000.
Cato. *De agricultura*. Trans. W. Hooper. *On Agriculture. LOEB Classical Library*, 283. Cambridge, MA: Harvard University Press, 1934.
Celsus. *De medicina*. Trans. W. G. Spencer. *On Medicine. LOEB Classical Library*, 292. Cambridge, MA: Harvard University Press, 1935.
Chrysostom, John. *Homily on I Cor*. Ed. Philip Schaff. *Early Church Fathers: Nicene-Post-Nicene, Series 1*, v. 12. www.ccel.org.
———. *In Act. Ap*. Ed. Philip Schaff. "On the Acts of the Apostles." *Early Church Fathers: Nicene-Post-Nicene Fathers, Series 1*, v. 11. www.ccel.org.
Cicero. *De leg*. Trans. Clinton W. Keyes. *On the Laws. LOEB Classical Library*. Cambridge, MA: Harvard University Press, 1928.
———. *In Catilinam Oratio*. Trans. C. McDonald. *Against Catiline. LOEB Classical Library*. Cambridge, MA: Harvard University Press, 1976.
Claudian. *Carmina minora, epithalamium*. Trans. M. Platnauer. *Wedding Poem of Honorius and Maria*. LOEB Classical Library. Cambridge, MA: Harvard University Press, 1956.
Clement of Alexandria. *Paedagogus*. Trans. Rev. Alexander Roberts and James Donaldson. *The Instructor. Early Church Fathers: Ante-Nicene Fathers*, v. 2. www.ccel.org.
———. *Stromata*. Trans. Rev. Alexander Roberts and James Donaldson. *Miscellaneous. Early Church Fathers: Ante-Nicene Fathers*, v. 2. www.ccel.org.
Columella. *De rustica*. Translated by E. S. Forster and Edward H. Heffner. *On Agriculture. LOEB Classical Library, 361, 408*. Cambridge, MA: Harvard University Press, 1941/1955.
Corripus. *In laudem iustini augusti minoris*. Trans. Averil Cameron. *In Praise of the Younger Justin*. London: Athlone Press, 1976.
Cyprian. *De dominica oratione. CSEL* 3.1.
———. *Epistles*. Ed. Philip Schaff. *Early Church Fathers: Ante-Nicene Fathers*, v. 5. www.ccel.org.
Early Arabic Poetry. V. 2. Trans. Alan Jones. Oxford: Ithaca Press Reading for the Board of the Faculty of Oriental Studies, 1996.
Eusebius. *History of the Church*. Trans. Paul L. Maier. Grand Rapids, MI: Kregel Academic, 2007.
Galen. *De alimentorum facultatibus*. Trans. Mark Grant. *On the Powers of Food* in *Galen on Food and Diet*. New York: Routledge, 2000.
———. *De optimo medico cognoscendo. Fī miḥnat afḍal al-aṭibbā'*. Trans. (from Arabic) Albert Z. Iskandar. *On Examinations by Which the Best Physicians Are Recognized. Corpus Medicorum Graecorum Supplementum Orientale*, 4. Berlin: Akademie-Verlag, 1988.

———. *De subtiliante diaeta. On the Thinning Diet*. Trans. P. N. Singer. *Selected Works*. New York: Oxford University Press, 1997.
al-Ghazālī, Abū Ḥāmid Muḥammad. *Iḥyā' `ulūm al-dīn*. Cairo, 1347 A.H.
———. *Kitāb dhikr al-mawt wa-mā ba`dahu*. Book XL of *Iḥyā' `ulūm al-dīn*. Trans. T. J. Winter. *The Remembrance of Death and the Afterlife*. Cambridge, UK: Islamic Texts Society, 1989.
———. *Kitāb al-durra al-fākhira fī kashf `ulūm al-ākhira*. Partial edition and French translation by L. Gautier, as *al-Dourra al-Fakhira: La Perle Précieuse de ghazali. Traité d'Eschatologie Musulmane*. Leipzig, 1877. English trans. Jane I. Smith, *The Precious Pearl*. MT: Scholars Press, 1979.
———. *Kitāb al-maḥabba wa'l-shawq wa'l-uns wa'l-riḍā*. Book XXXVI of *Iḥyā' `ulūm al-dīn*. Trans. Eric Ormsby. *Book of Love, Longing, Intimacy, and Contentment*. Cambridge, UK: Islamic Texts Society, 2011.
Gregory of Tours. *Gloria confessorum*. Edited by Bruno Krusch. *MGH SRM* I.2. Trans. Raymond Van Dam, *Gregory of Tours, Glory of the Confessors*. Liverpool, UK: Liverpool University Press, 1988.
———. *In gloria martyrum*. Edited by Bruno Krusch. *MGH SRM* 1.2. Trans. Raymond Van Dam, *Gregory of Tours, Glory of the Martyrs*. Liverpool: Liverpool University Press, 1988.
———. *Wonders of St. Martin*. Trans. in *Monks, Bishops, and Pagans*, ed. E. Peters. Philadelphia: University of Pennsylvania Press, 1975
Gregory the Great. *Liber regulae pastoralis*. *SC* nos. 381–82. Trans. George E. Demacopoulos. *The Book of Pastoral Rule*. New York: St. Vladimir's Seminary Press, 2007.
Al-Harawī, `Alī ibn Abī Bakr. *Kitāb al-ishārāt ilā ma`rifat al-ziyārāt*. Trans. (English) Josef W. Meri, *A Lonely Wayfarer's Guide to Pilgrimage*. Princeton, NJ: Darwin Press, Inc., 2004. Trans. (French) Janine Sourdel-Thomine, *Guide Des Lieux De Pèlerinage*, Damas: Institu Français de Damas, 1957.
Hegel. *Aesthetics*. Trans. T. M. Knox. Oxford: Oxford University Press, 1975.
Herodian. *Histories*. Trans. C. R. Whittaker. *LOEB Classical Library*, 2 vols. Cambridge, MA: Harvard University Press, 1969.
Herodotus. *Histories*. Trans. George Rawlinson. Knoxville, TN: Wordsworth Classics, 1996.
Horace. *Carmina*. Trans. David Ferry. *Odes of Horace*. New York: Farrar, Straus, and Giroux, 1997.
———. *Satirae*. Trans. Rushton Fairclough. *Satires, Epistles, and Ars Poetica. LOEB Classical Library*. Cambridge, MA: Harvard University Press, 1929.
Ibn Abī Shayba. *Musannaf*. Ed. Shâhin. Beirut: 1995.
Ibn al-Arabi. *Fuṣūṣ al-ḥikam*. Trans. R. W. J. Austin. *The Bezels of Wisdom*. Mahwah, NJ: Paulist Press, 1980.
Ibn Baṭṭūṭa. *The Travels of Ibn Baṭṭūṭa*. Ed. Tim Mackintosh-Smith. London: Pan Macmillan, 2003.
Ibn Ḥajar al-`Asqalānī, Aḥmad b. `Alī. *Fatḥ al-bārī bi-sharḥ ṣaḥīḥ al-bukhārī*. Beirut: Dār Iḥyā al-Turāth al-`Arabī, 1988.
Ibn al-Ḥājj, Muḥammad b. Muḥammad. *Madkhal al-shar` al-sharīf*. 4 vols. Cairo, 1929.
Ibn Hanbal. *Musnad*. Cairo, 1313/1896.
Ibn Ishaq. *The Life of Muhammad: A Translation of Isḥāq's Sīrat rasūl allāh*. Trans. A. Guillaume. Oxford: Oxford University Press, 1997.
Ibn Kathir. *Al-sīra al-nabawiyya*. Translated and published separately from his *Al-bidāya wa al-nihāya*, Trans. Trevor Le Gassick, Reviewed Ahmed Fareed. UK: Garner Publishing Limited, 1988.
———. *Tafsīr al-qur'ān al-`aẓīm*. Bayrūt: Dār al-andalus lil-ṭibā'ah wa-al-nashr, 1966.
Ibn Māsawaih. *Kitāb jawāhir al-ṭīb al-mufrada*. Ed. P. Sbath. "Traité sur les substances simples aromatiques." *Bulletin de l'Institut d'Égypte* 19 (1936–1937): 5–27.
———. Trans. Martin Levey. "On Simple Aromatic Substances." *Journal of the History of Medicine and Allied Sciences* 16.4 (1961): 394–410.
Ibn Taymiyya. *Kitāb iqtiḍā' al-sirāṭ al-mustaqīm mukhālafat aṣḥāb al-jaḥīm. Ibn Taimīya's Struggle against Popular Religion*. Trans. Muhammad Umar Memon. Paris: Mouton & Co., 1976.

Ignatius of Antioch. *Ad. Rom*. Ed. Philip Schaff. *Letter to the Romans. Early Church Fathers: Ante-Nicene Fathers*, v. 1 www.ccel.org.
Imru al-Qays. *Mu`allaqa*. Trans. Alan Jones. *Early Arabic Poetry*, v. 2. Oxford: Ithaca Press Reading for the Board of the Faculty of Oriental Studies, 1996.
Al-Iṣfahānī, Abū al-Faraj. *Kitāb al-aghānī*. Cairo, 1323/1905–1906.
Al-Isqalānī, Ibn Hajar. *Fatḥ al-bārī bi-sharḥ ṣaḥīḥ al-bukhārī*. Beirut: Dār Iḥyā al-Turāth al-`Arabī, 1988.
Jerome. *Epistulae. CSEL* 54–56. Trans. William H. Fremantle, George Lewis, and William G. Martley. *Letters. Early Church Fathers: Nicene and Post-Nicene Fathers, Second Series*, v. 6. www.ccel.org.
———. *Life of Saint Hilarion*. Eds. Philip Schaff and Allan Menzies. *Early Church Fathers: Nicene-and-Post-Nicene, Series 2*, v. 6. www.ccel.org.
Justin Martyr. *Apologia*. Ed. Philip Schaff. *Apology. Early Church Fathers: Ante-Nicene Fathers*, v. 1. www.ccel.org.
Juvenal. *Satirae*. Trans. Peter Green. *The 16 Satires*. Harmondsworth, UK: Penguin, 1967.
Kanz al-fawā'id fī tanwī al-mawā'id. Eds. Marín and Waines. Beirut: Franz Steiner, 1993.
al-Kindi, Ya'qūb b. Ishaq. *Aqrābādhīn*. Trans. Martin Levey. *The Medical Formulary*. Madison: University of Wisconsin Press, 1966.
———. *Kitāb fi-kīmiyā' al-itr wat-taṣ`īdāt*. Ed. and trans. Karl Garbers. *Abhandlungen für die Kunde des Morgenländes XXX*. Leipzig, 1948.
———. *Risālat al-kindī fī ajzā' khabariyyah fī al-mūsīqā*. Cairo: al-Lajnah al-Mūsīqiyyah al-`Ulyā, 1963.
Al-Kisa'i. *Qiṣaṣ al-anbiyā*. Trans. Wheeler M. Thackston. Chicago: Kazi Publications, 1997.
Kitāb al-hadāyā wa al-tuḥaf. Trans. Ghada al-Hijjawai al-Qaddami. *Books of Gifts and Rarities*. Cambridge, MA: Harvard University Press, 1996.
Lactantius. *Divinae Institutiones*. Trans. M. F. McDonald. *Divine Institutes. Early Church Fathers: Ante-Nicene Fathers*, v. 7. www.ccel.org.
Letter from Vienna and Lyon. From Eusebius, *History of the Church*, 5.1. Trans. Paul L. Maier. Grand Rapids, MI: Kregel Academic, 2007.
Liber Pontificalis. Trans. Raymond Davis. *The Book of Pontiffs: The Ancient Biographies of the First Ninety Roman Bishops to AD 710*. Liverpool: Liverpool University Press, 1989.
Life of Simeon Stylites: A Translation of the Syriac Text in Bedjan's Acta Martyrum et Sanctorum, Vol. IV. Trans. Frederick Lent. *Journal of the American Oriental Society* 35 (1915): 103–11.
Lucian. *Amores*. Trans. M. D. MacLeod. *LOEB Classical Library*, 8 vols. Cambridge, MA: Harvard University, 1967.
———. *Icaromenippus*. Trans. C. D. N. Costa. *High Above the Clouds. Selected Dialogues*. New York: Oxford University Press, 2005.
Lucretius. *De rerum natura*. Trans. Martin Ferguson Smith. *On the Nature of Things*. Cambridge, MA: Hackett Publishing Company, Inc., 2001.
Al-Majlisī. *Biḥār al-anwār*. Tehran-Qum, 1376–1392/1956–1972.
Martial. *Epigrammata*. Trans. D. R. Shackleton Bailey. *Epigrams. LOEB Classical Library*, 3 vols. Cambridge, MA: Harvard University, 1993.
The Martyrdom of Ignatius. Eds. Alexander Roberts and James Donaldson. *Early Church Fathers: Ante-Nicene Fathers*, v. 1. www.ccel.org.
The Martyrdom of Perpetua and Felicitas. Ed. Allan Menzies. *Early Church Fathers: Ante-Nicene Fathers*, v. 3. www.ccel.org.
The Martyrdom of Polycarp. Eds. Alexander Roberts and James Donaldson. *Early Church Fathers: Ante-Nicene Fathers*, v. 1. www.ccel.org.
The Martyrdom of the Holy Martyrs Justin, Chariton, Charites, Paeon, Liberianus, Who Suffered at Rome. Eds. Alexander Roberts and James Donaldson. *Early Church Fathers: Ante-Nicene Fathers*, v. 1. www.ccel.org.
"Missale gothicum": A Gallican sacramentary, ms. Vatican. regin. lat. 317. Edited, with introduction, diplomatic, and liturgical notes. Reprints from University of Michigan Library, 1917.
Al-Mutanabbī, Abū al-Ṭayyib Aḥmad b. al-Ḥusayn. *Dīwān*. Beirut: Dār Ṣādir, 1958.

Novatian. *De spectabilis*. Trans. Russell J. DeSimone. *The Spectacle*. Washington, DC: Catholic University of America Press, 1974.
The Old Testament Pseudepigrapha. Ed. James H. Charlesworth. New York: Doubleday, 1983–1985.
Origen. *Contra celsum*. PG 11. Ed. A. Cleveland Coxe. *Early Church Fathers: Ante-Nicene Fathers*, v. 4. www.ccel.org.
———. *De oratione*. *PG* 11. Trans. William Curtis. *Early Church Fathers: Ante-Nicene Fathers*, v. 4. www.ccel.org.
———. *Dialogue with Herclides*. Trans. Robert J. Daley. *Ancient Christian Writers*, 54. Nahwah, NJ: Paulist Press, 1992.
———. *Exhortatio*. Trans. Rowan A. Greer. *Exhortation to Martyrdom: Classics of Western Spirituality*. Mahwah, NJ: Paulist Press, 1979.
Ovid. *Ars amatoria*. Trans. J. H. Mozley. *The Art of Love and Other Poems. LOEB Classical Library*, 6 vols. Cambridge, MA: Harvard University Press, 1979.
———. *Ex ponto*. Trans. A. L. Wheeler, revised by G. P. Goold. *Letters from the Black Sea. LOEB* Classical Library, 151. Cambridge, MA: Harvard University Press, 1924.
———. *Fasti*. Trans. James G. Frazer, revised by G. P. Goold. *The Book of Days. LOEB Classical Library*, 253. Cambridge, MA: Harvard University Press, 1931.
———. *Metamorphosis*. Trans. Frank Justus Miller, revised by G. P. Goold, *LOEB Classical Library*, v. 42. Cambridge, MA: Harvard University Press, 1916. www.ccel.org.
Paulinus of Nola. *Poems*. Trans. P. G. Walsh. *Ancient Christian Writers*, v. 40. Mahwah, NJ: Paulist Press, 1974.
Petronius. *Satyricon*. Trans. Sarah Ruden. Indianapolis: Hackett Publishing Company, Inc., 2000.
Philo. *On the Special Laws*. Trans. F. H. Colson. *LOEB Classical Library*, 320. Cambridge, MA: Harvard University Press, 1937.
Plautus. *Aulularia*. Trans. Palmer Bovie. *Pot of Gold. The Comedies, V. II*. Eds. David R. Slavitt and Palmer Bovie. Baltimore: Johns Hopkins University Press, 1995.
Pliny the Elder. *Naturalis historia*. Trans. H. Rachham. *Natural History. LOEB Classical Library*, 10 vols. Cambridge, MA: Harvard University Press, 1938–1962.
Prudentius, *A Hymn on the Burial of the Dead*. Trans. H. J. Thomson. Cambridge/London: 1949.
Al-Qadi, 'Abd al-Raḥīm ibn Aḥmad. *Daqā'iq al-akhbār fī dhikr al-jannah wa-al-nār*. Trans. A'isha 'Abd ar-Rahman at-Tarjumana. *Islamic Book of the Dead: A Collection of Hadiths on the Fire and the Garden*. San Francisco: Diwan Press, 1977.
Al-Raffā, al-Sarī. *Al-Muḥibb wa al-maḥbūb wa al-mashmūm wa al-mashrūb*, ed. Miṣbāh. Ghalāwinjī (Damascus, 1986–87).
Al-Rāzī, Abū Bakr. *Ḥāwī: Kītāb al-ḥawī fī al-ṭibb*. Haydarabad: Osmania Oriental Publications Bureau, 1960.
Sābūr ibn Sahl. *Dispensatorium parvum (al-Agrābādhin al-ṣaghīr)*. Analyzed, edited, and annotated by Oliver Kahl. New York: Brill, 1994.
———. *The Small Dispensatory*. Trans. Oliver Kahl. Boston: Brill, 2003.
———. *Dispensatory in the Recension of the `Aḍudī Hospital*. Trans. Oliver Kahl. Boston: Brill, 2009.
Al-Ṣan`ānī, Abd al-Razzāq. *Muṣannaf*. Bayrūt: al-Majlis al-`Ilmī, 1970.
Semahot. The Tractate "Mourning": Regulations Relating to Death, Burial, and Mourning. Trans. Dov Zlotnick. New Haven, CT: Yale University Press, 1966.
Seneca. *Epistulae morales ad lucilium*. Trans. Richard M. Gummere. *Moral Letters to Lucilius. LOEB Classical Library*, 3 vols. Cambridge, MA: Harvard University Press, 1917–1925.
The Seven Odes: The First Chapter in Arabic Literature. Trans. A. J. Arberry. New York: Macmillan Co., 1957.
Al-Sha`rānī. *Mukhṭaṣar al-tadhkira al-qurṭubīyya*. Cairo: n.d.
Sibṭ Ibn al-Jawzī, *Mir'āt al-zamān fī ta'rīkh al-ayān*. Hyderabad: 1951–1952.
Suetonius. *Lives of the Later Caesars*. Trans. J. C. Rolfe. *LOEB Classical Library*, 31. Cambridge, MA: Harvard University Press, 1964.

Sulpicius Severus. *Vita s. martini*. Latin text with French translation by Jacques Fontaine, *SC* 133. English translation by F. R. Hoare, "The Life of Saint Martin of Tours." In *Soldiers of Christ: Saints and Saints' Lives from Late Antiquity and the Early Middle Ages*, eds. Thomas F. X. Noble and Thomas Head. University Park: Pennsylvania State Press, 1995.

al-Suyūṭī, Jalal al-Din. *al-Amr bi-al-ittibā wa al-nahy `an al-ibtidā`*. Ed. M. H. Salmān. Al-Dammām, 1990.

———. *Angels in Islam: Jalal al-Din al-Suyūṭī's al-Haba'ik fi akhbar al-mala'ik*. Trans. Stephen Burge. New York: Routledge, 2011.

———. *Al-Haba'ik fi akhbar al-mala'ik*. Trans. Stephen Burge. In *Angels in Islam: Jalal al-Din al-Suyūṭī's al-Haba'ik fi akhbar al-mala'ik*. New York: Routledge, 2011.

———. *al-Maqama al-miskiyya*. Trans. G. J. H. Van Gelder. "Four Perfumes of Arabia." *In Parfums d'Orient. Textes reunis par R. Gyselen*. Bures-sur-Yvette: Groupe pour l'étude de la civilization du moyen-orient, 1998.

———. *Ṭibb al-Nabi*. Sidney, NY: American Institute of Tibb, 1985. Trans. Cyril Elgood. *Osiris* V. 14, 1962.

The Syriac Book of Medicines: Syrian Anatomy, Pathology and Therapeutics in the Early Middle Ages, with Sections on Astrological and Native Medicine and Recipes. Ed. and trans. Ernest A. Wallis Budge. St. Helier: Armorica Book Co., 1913.

The Syriac Chronicle Known as That of Zachariah of Mitylene. Trans. Frederick John Hamilton. Nabu Press, Reproduction Series, 2010 (1899).

al-Ṭabarī, Abū Ja`far Muḥammad ibn Jarīr. *Tafsīr al-Ṭabarī: Jāmi' al-bayān 'an ta'wīl al-Qur'ān*. Cairo: Dār al-Ma'ārif, 1955.

———. *Ta'rīkh al-rasul wa'l-mulūk*. Trans. Franz Rosenthal. *The History of al-Ṭabarī*. Albany: State University of New York Press, 1989.

al-Ṭabarī, `Alī b. Sahl Rabban. *Firdaws al-ḥikmah fī al-tibb*. Ed. M. Z. Siddiqi. Berlin: Sonne, 1928.

Al-Tayalisi, Sulayman ibn Dawud. *Musnad*. Bayrut: Dar-al-Maa'rifah, 1999.

Al-Tha`alabī, Aḥmad ibn Muḥammad. *Arais al-majalis fi qisas al-anbiya*. Trans. William M. Brinner. *Lives of the Prophets*. London: Brill, 2002.

Al-Tirmidhi. *Shama'il*. Trans. (with Arabic) Muhammad bin Abdur Rahman Ebrahim. New Delhi: Adam Publishers & Distributors, 2012.

Tertullian. *Ad martyras*. Ed. Allan Mezies. *To the Martyrs. Fathers of the Church: Ante-Nicene Fathers*, v. 3. www.ccel.org.

———. *Apologia*. Ed. A. Cleveland Coxe. *Apology. Fathers of the Church: Ante-Nicene Fathers*, v. 4 www.ccel.org.

———. *De cultu feminarum. SC* 173. Ed. A. Cleveland Coxe. *On the Apparel of Women. Fathers of the Church: Ante-Nicene Fathers*, v. 4. www.ccel.org.

———. *De idol. CSEL* 20. Ed. A. Cleveland Coxe. *Concerning Idolatry. Fathers of the Church: Ante-Nicene Fathers*, v. 4. www.ccel.org.

———. *De ieiunio*. Ed. A. Cleveland Coxe. *On Fasting. Fathers of the Church: Ante-Nicene Fathers*, v. 4. www.ccel.org.

———. *De oratione*. Ed. Allan Mezies. *Concerning Prayer. Fathers of the Church: Ante-Nicene Fathers*, v. 3. www.ccel.org.

———. *De spectaculis*. Ed. Allan Mezies. *On Shows. Fathers of the Church: Ante-Nicene Fathers*, v. 3. www.ccel.org.

Theodoret of Cyrrus. *SC* 234, 257. *A History of the Monks of Syria*. Trans. and intro. R. M. Price. Kentucky: Cistercian Press, 1985.

Theophrastus. *De odoribus*. Trans. Arthur Hort. *On Odors. LOEB Classical Library*. Cambridge, MA: Harvard University Press, 1916.

Tibullus, *Carmina. Oxford Classical Texts*. Oxford: Clarendon Press, 1993.

Usāmah ibn Munqidh. *The Book of Contemplation. Islam and the Crusades*. Trans. and intro. Paul M. Cobb. New York: Penguin, 2008.

Vergil. *Aeneid*. Trans. Robert Fitzgerald. New York: Vintage Classics, 1990.

Vita genovefae. MGH.SRM 3.204–38. Trans. Jo Ann McNamara, John E. Halborg, with E. Gordon Whatley. *Sainted Women of the Dark Ages*. Durham, NC: Duke University Press, 1992.

Vita radegundis. *MGH.SRM* 2.358–405. Trans. Jo Ann McNamara, John E. Halborg, with E. Gordon Whatley. *Sainted Women of the Dark Ages*. Durham, NC: Duke University Press, 1992.

Al-Warrāq, Ibn Sayyār. Trans. Nawal Nasrallah. *Annals of the Caliphs' Kitchens*. Boston: Brill, 2007.

Zeno of Verona. *Sermons*. Trans. Gordon P. Jeanes. In *The Day Has Come! Easter and Baptism in Zeno of Verona*. Collegeville, MN: Liturgical Press, 1995.

———. *Tractatus*. *CCSL* 22.

SECONDARY SOURCES

Abrahamov, Binyamin. "The Creation and Duration of *Paradise* and Hell in Islamic Theology," *Der Islam* 79.1 (2002): 87–102.

Aho, James. *The Orifice as Sacrificial Site: Culture, Organization, and the Body*. Dulles, VA: AldineTransaction, 2002.

Ahsan, Muhammad Munazir. *Social Life Under the `Abbasids*. New York: Longman Group, Ltd., 1979.

Alazmeh, A. "Rhetoric for the Senses: A Consideration of Muslim Paradise Narratives." *Journal of Arabic Literature* 26.3 (1995): 215–31.

Albert, Jean-Pierre. *Odeurs de Sainteté: La Mythologie Chrétienne des Aromates*. Paris: École des Hautes Études en Sciences Sociales. 1996.

Allen, Valerie. *On Farting: Language and Laughter in the Middle Ages*. New York: Palgrave Macmillan, 2007.

Amir-Moezzi, Mohammad Ali. *The Divine Guide in Early Shi`ism: The Sources of Esotericism in Islam*. Albany: State University of New York Press, 1994.

Anderson, Eugene. *Everyone Eats: Understanding Food and Culture*. New York: New York University Press, 2005.

Anderson, Toby. "The Arab Influence on Western European Cooking." *Journal of Medieval History* 6 (1980): 317–40.

Atchley, E. G. Cuthbert F. *A History of the Use of Incense in Divine Worship*. Charleston: Nabu Press, 2010.

Attridge, Harold W. "Making Scents of Paul: The Background and Sense of 2 Cor. 2:14–17." *Early Christianity and Classical Culture: Comparative Studies in Honor of Abraham J. Malherbe*. Ed. John T. Fitzgerald. Leiden: Brill, 2003.

Avrahami, Yael. *The Senses of Scripture: Sensory Perception in the Hebrew Bible*. New York: T&T Clark, 2012.

Baader, Gerhard. "Early Medieval Latin Adaptations of Byzantine Medicine in Western Europe." *Dumbarton Oaks Papers*. V. 38, Symposium on Byzantine Medicine (1984): 251–59.

Barton, Carlin A. "Savage Miracles: The Redemption of Lost Honor in Roman Society and the Sacrament of the Gladiator and the Martyr." *Representations* 45 (Winter 1994): 41–71.

Baskin, Judith. *Midrashic Women: Formations of the Feminine in Rabbinic Literature*. Hanover: Brandeis University Press, 2002.

Beard, M. "Imaginary *Horti*: Or Up the Garden Path." In *Horti Romani*, eds. M. Cima and E. LaRocca. Rome: L'Erma di Bretschneider, 1998.

Beer, Michael. *Taste or Taboo: Dietary Choices in Antiquity*. Devon: Prospect Books, 2010.

Benkheira, Mohammad Hocine. *"La maison de Satan:* La hammâ en débat dans l'islam medieval.*" Revu de l'histoire des religions*, 200.4 (2003): 391–443; 224.3 (2007): 319–71; 225.1 (2008): 75–128.

Bergmann, B., and C. Kondoleon, eds. *The Art of Ancient Spectacle*. New Haven: Yale University Press, 1999.

Berkey, Jonathan. *Popular Preaching and Religious Authority in the Islamic Medieval Near East*. Seattle: University of Washington Press, 2001.

Bickerman, E. *Le Culte des Souverains dans Souverains dans l'Empire Romain*. Vandoeuvres-Genève: Fondation Hardt, 1973.

Bockmuehl, Markus. "Locating Paradise." In *Paradise in Antiquity*, eds. Markus Bockmuehl and Guy G. Stroumsa. New York: Cambridge University Press, 2010.
Bonnet, M. *Le Latin de Grégoire de Tours*. Paris: Librairie Hatchette, 1890.
Bos, Gerrit. "Ibn al-Jazzār on Medicine for the Poor and Destitute." *Journal of the American Oriental Society* 118.3 (July–September 1998): 365–75.
Bouillart, Jacques. *Histoire de l'abbeye royale de Saint Germain des Pres*. Paris: 1724.
Boustan, Ra`anan. *Heavenly Realms and Earthly Realities in Late Antique Religions*. Cambridge, UK: Cambridge University Press, 2004.
Boyer, R. *Le sacrophage à sa découverte, in Premiers temps chrétiénes en Gaule méridionale: antìquité tardive et haut moyen âge, IIIème–VIIIème siècles, Catalogue de l'exposition.* Lyons, 1986.
Bradley, Mark, ed. *Smell and the Ancient Senses*. New York: Routledge, 2015.
Bradshaw, Paul. *The Search for the Origins of Christian Worship: Sources and Methods for the Study of Early Liturgy*. London: SPCK, 1992.
Brand, Yehoshua. *Keley zekhukhit besifrut ha-talmud*. Jerusalem: Mosad Ha-Rav Koon, 1978.
Bremmer, Jan N. "Paradise: From Persia, Via Greece, into the *Septuagint*." In *Paradise Interpreted: Representations of Biblical Paradise in Judaism and Christianity*, ed. Gerard P. Luttikhuizen. Leiden: Brill, 1999.
Brent, Allen. *Cyprian and Roman Carthage*. New York: Cambridge University Press, 2010.
———. *The Imperial Cult and the Development of Church Order: Concepts and Images of Authority in Paganism and Early Christianity Before the Age of Cyprian*. Leiden: Brill, 1999.
Brett, Gerard. "The *Automata* in the Byzantine 'Throne of Solomon,'" *Speculum* XXIX.3 (1954): 477–87.
Brinner, William M. "Prophet and Saint: The Two Exemplars of Islam." In *Saints and Virtues*, ed. John Stratton Hawley. Berkeley: University of California Press, 1987.
———. English translation, Aḥmad ibn Muḥammad Al-Tha`alabī. *Arais al-majalis fi qisas al-anbiya/Lives of the Prophets*. London: Brill, 2002
Brockopp, Jonathan E., ed. *The Cambridge Companion to Muhammad*. Cambridge, UK: Cambridge University Press, 2010.
Brown, Peter. *The Body and Society: Men, Women and Sexual Renunciation in Early Christianity*. New York: Columbia University Press, 1988.
Bruno, Laurioux. *Les Livres de cuisine médiévaux*. (Typologies des Sources du Moyen Age Occidental 77.) Belgium: Brepols, 1997.
Bulliet, R. W. *The Camel and the Wheel*. Cambridge, MA: Harvard University Press, 1974.
Burge, S. R. "Impurity/Danger!" *Islamic Law and Society* 17 (2010): 320–49.
———. "Angels, Ritual and Sacred Space in Islam." *Comparative Islamic Studies* 5.2 (2009): 221–45.
Burn, Jean-Pierre. "The Production of Perfumes in Antiquity: The Cases of Delos and Paestum." *American Journal of Archaeology* 104.2 (April 2000): 277–308.
Burr, Chandler. *The Emperor of Scent: A Story of Perfume, Obsession, and the Last Mystery of the Senses*. New York: Random House, 2002.
Burton, John. "The Qur'an and the Islamic Practice of *Wuḍū*." *BSOAS* 51 (1988): 21–58.
Campo, Juan Eduardo. *The Other Sides of Paradise: Explorations into the Religious Meanings of Domestic Space in Islam*. Columbia: South Carolina Press, 1991.
Carlisle, Janice. *Common Scents: Comparative Encounters in High-Victorian Fiction*. Oxford: Oxford University Press, 2004.
Caseau, Beatrice. *Euodia: The Use and Meaning of Fragrances in the Ancient World and their Christianization (100–900)*. Unpublished dissertation, 1994.
Chernus, Ira. "Visions of God in Merkabah Mysticism." *Journal for the Study of Judaism* XIII.1–2 (1982): 123–46.
Clancy SJ, Finbarr G. "Imitating the Mysteries That You Celebrate: Martyrdom and Eucharist in the Early Patristic Period." In *The Great Persecution: The Proceedings of the Fifth Patristic Conference, Maynooth, 2003*, eds. Vincent Twomey SVD and Mark Humphries. Dublin: Four Courts Press, 2009, 106–40.

Classen, Constance, David Howes, and Anthony Synnott, eds. *Aroma: The Cultural History of Smell*. New York: Routledge, 1994.
Conan, M. "Nature into Art: Gardens and Landscapes in the Everyday Life of Ancient Rome." *Journal of Garden History* 6.4 (1986): 348–56.
Connors, Catherine. "Scents and Sensibility in Plautus' *Casina*." *The Classical Quarterly, New Series* 47.1 (1997): 305–9.
Cook, David. *Studies in Muslim Apocalyptic*. Princeton, NJ: Darwin Press, 2002.
Coolman, Boyd Taylor. *Knowing God by Experience: The Spiritual Senses in the Theology of William of Auxerre*. Washington, DC: Catholic University of America Press, 2004.
Corbin, Alain. *The Foul and the Fragrant: Odor and the French Social Imagination*. Cambridge, MA: Harvard University Press, 1986.
Cormack, Margaret, ed. *Muslims and Others in Sacred Space*. Oxford: Oxford University Press, 2013.
Corn, Charles. *The Scents of Eden: A Narrative of the Spice Trade*. New York: Kodansha International, 1998.
Cramer, Peter. *Baptism and Change in the Early Middle Ages, c. 200–1150*. New York: Cambridge University Press, 1993.
Crislip, Andrew T. *From Monastery to Hospital: Christian Monasticism and the Transformation of Health Care in Late Antiquity*. Ann Arbor: University of Michigan Press, 2005.
Crone, Patricia. *Meccan Trade and the Rise of Islam*. Princeton, NJ: Princeton University Press, 1987.
Curta, Florin. "Merovingian and Carolingian Gift Giving." *Speculum* 81 (2006): 671–99.
D'Arms, John H. "The Culinary Reality of Roman Upper-Class *Convivia*: Integrating Texts and Images." *Comparative Studies in Society and History* 46.3 (July 2004): 428–50.
———. "Performing Culture: Roman Spectacle and the Banquets of the Powerful." In *The Art of Ancient Spectacle*, eds. B. Bergmann and C. Kondoleon. New Haven: Yale University Press, 1999, 301–19.
Davidson, Clifford, ed. *The Iconography of Heaven*. Kalamazoo, MI: Medieval Institute Publications, 1994.
Dayagi-Mendels, Michal. *Perfumes and Cosmetics in the Ancient World*. Jerusalem: Israel Museum, 1993.
DeConick, April D., ed. *Paradise Now: Essays on Early Jewish and Christian Mysticism*. Atlanta: Society of Biblical Literature, 2006.
Dench, E. "Austerity, Excess, Success, and Failure in Hellenistic and Early Imperial Italy." In *Parchments of Gender*. Oxford: Clarendon Press, 1998, 121–46.
Denny, Frederick. "God's Friends: The Sanctity of Persons in Islam." In *Sainthood: Its Manifestation in World Religions*, eds. Richard Kieckhefer and George Doherty Bond. Berkeley: University of California Press, 1988, 69–97
Detienne, Marcel. *Gardens of Adonis: Spices in Greek Mythology*. Princeton, NJ: Princeton University Press, 1994.
Dols, Michael. "Insanity in Byzantine and Islamic Medicine." *Dumbarton Oaks Papers* 38 (1984): 135–48.
Dorland, Gabrielle J. *Scents Appeal: The Silent Persuasion of Aromatic Encounters*. New Jersey: Wayne Dorland Company, 1993.
Duchesne, L. *Christian Worship: Its Origins and Evolution, 5th ed.* Trans. M. L. McClure. New York: Macmillan Co., 1931.
Dunbabin, K. "Triclinium and Stibadium." In *Dining in a Classical Context*, ed. William J. Slater. Ann Arbor: University of Michigan Press, 1991, 121–48.
Dvorjetski, Estēe. *Leisure, Pleasure and Healing: Spa Culture and Medicine in Ancient Eastern Mediterranean*. Boston: Brill, 2007.
Edwards, Catherine. *Death in Ancient Rome*. New Haven: Yale University Press, 2007.
Effros, Bonnie. *Caring for the Body and Soul: Burial and the Afterlife in the Merovingian World*. University Park: Pennsylvania State University Press, 2008.
———. *Creating Community with Food and Drink in Merovingian Gaul*. New York: Palgrave Macmillan, 2002.

Eliav, Yaron Z. "The Roman Bath as a Jewish Institution: Another Look at the Encounter between Judaism and the Greco-Roman Culture." *Journal for the Study of Judaism* 31.4 (2000): 416–54.
Engen, Trygg. *Odor Sensation and Memory*. New York: Praeger, 1991.
Evans, Suzanne. "The Scent of a Martyr." *Numen: International Review for the History of Religions* 49.2 (2002): 193–211.
Ezz El-Din, Hassan El-Banna. "No Solace for the Heart: The Motif of the Departing Women in the Pre-Islamic Battle Ode." In *Reorientations/Arabic and Persian Poetry*, ed. Suzan Pinckney Stetkevych. Bloomington: Indiana University Press, 1994.
Fagan, Garrett G. "Bathing for Health with Celsus and Pliny the Elder." *Classical Quarterly* 56.1 (2006): 190–207.
———. *Bathing in Public in the Roman World*. Ann Arbor: University of Michigan Press, 2002.
Farooqi, M. I. H. *Medicinal Plants in the Traditions of Prophet Muhammad: Medicinal, Aromatic and Food Plants Mentioned in the Traditions of Prophet Muhammad*. Lucknow: Sidrah Publishers, 1998.
Feliks, Yehuda. "The Incense of the Tabernacle." In *Pomegranates and Golden Bells: Studies in Biblical, Jewish, and Near Eastern Ritual, Law, and Literature in Honor of Jacob Milgrom*, eds. David P. Wright, David Noel Freedman, and Avi Hurvitz. Winona Lake, IN: J. E. Eisenbrauns, 1995.
Ferngren, Gary B. *Medicine and Health Care in Early Christianity*. Baltimore: Johns Hopkins University Press, 2009.
Fierro, M. "The Treatises against Innovation *(Kutub al-bida)*." *Der Islam* 69 (1992): 204–46.
Fishwick, Duncan. "Prudentius and the Cult of Divus Augustus." *Historia: Zeitschrift für Alte Geschichte* 39.4 (1990): 475–86.
Flemming, Rebecca. *Medicine and the Making of Roman Women: Gender, Nature, and Authority from Celsus to Galen*. Oxford: Oxford University Press, 2000.
Fonrobert, Charlotte Elisheva. *Menstrual Purity: Rabbinic and Christian Reconstructions of Biblical Gender*. Stanford: Stanford University Press, 2000.
Frank, Georgia. *The Memory of the Eyes: Pilgrims to Living Saints in Christian Late Antiquity*. Berkeley: University of California Press, 2000.
———. "'Taste and See': The Eucharist and the Eyes of Faith in the Fourth Century." *Church History* 70.4 (December 2001): 619–43.
Fredrick, David. *The Roman Gaze*. Ed. David Fredrick. Baltimore: John Hopkins University Press, 2002.
Freidenreich, David M. *Foreigners and Their Food: Constructing Otherness in Jewish, Christian, and Islamic Law*. Berkeley: University of California Press, 2011.
Fulton, Rachel. "'Taste and See that the Lord is Sweet' (Ps. 33:9): The Flavor of God in the Monastic West." *Journal of Religion* 86.2 (2006): 169–204.
Gardiner, Eileen, ed. *Visions of Heaven and Hell Before Dante*. NY: Italica Press, 1989.
Giesecke, Annette. *The Epic City: Urbanism, Utopia, and the Garden in Ancient Greece and Rome*. Cambridge, MA: Harvard University Press, 2007.
Giordano, C., and A. Casale. *Profumi, Unguenti e acconciature in Pompei antica*. Rome: Bardi, 1992.
Goitein, S. D. "Portrait of a Medieval India Trader: Three Letters from the Cairo Geniza." *Bulletin of the School of Oriental and African Studies* 50.3 (1987): 449–64.
Goodman, Martin. "Paradise, Gardens and the Afterlife in the First Century CE." In *Paradise in Antiquity: Jewish and Christian Views*, eds. Markus Bockmuehl and Guy G. Strouma. Cambridge, UK: Cambridge University Press, 2010.
———. *State and Society in Roman Galilee, A.D. 132–212*. Totowa, NJ: Rowman & Allanheld, 1983.
Goody, Jack. *Cooking, Cuisine and Class*. Cambridge, UK: Cambridge University Press, 1982.
Gradel, Ittai. *Emperor Worship and Roman Religion*. Oxford: Clarendon Press, 2002.
Green, Deborah. *The Aroma of Righteousness: Scent and Seduction in Rabbinic Life and Literature*. University Park: Pennsylvania State University Press, 2011.

Grimm, Veronika E. *From Feasting to Fasting: The Evolution of a Sin: Attitudes to Food in Late Antiquity*. New York: Routledge, 1996.
Grotzfeld, Heinz. *Das Bad in arabisch-islamischen Mittelalter*. Eine Kulturgeschichtl. Wiesbaden: Harrassowitz, 1970.
Groom, Nigel. *Frankincense and Myrrh: A Study of the Arabian Incense Trade*. Harlow, Essex: Longman Group Limited, 1981.
———. "Trade, Incense, and Perfume." In *Caravan Kingdoms: Yemen and the Ancient Incense Trade*, ed. Ann C. Gunter. Washington, DC: Smithsonian Institution, 2005, 105–13.
Gunter, Ann Clyburn. *Caravan Kingdoms: Yemen and the Ancient Incense Trade*. Washington, DC: Smithsonian Institution, 2005.
Guy, Laurie. "'Naked' Baptism in the Early Church: The Rhetoric and the Reality." *Journal of Religious History* 27.2 (June 2003): 133–42.
Hale, Rosemary Drage. "'Taste and see, for God is Sweet': Sensory Perception and Memory in Medieval Christian Mystical Experience." In *Vox Mystica: Essays on Medieval Mysticism in Honor of Professor Valerie M. Lagorio*, ed. Anne Clark Bartlett. Rochester: D. S. Brewer, 1995.
Halevi, Leor. *Muhammad's Grave: Death Rites and the Making of Islamic Society*. New York: Columbia University Press, 2007.
Haran, Menahem. "The Uses of Incense in the Ancient Israelite Ritual." *Vetus Testamentum* 10.2 (April 1960): 113–29.
Harrill, J. Albert. "Coming of Age and Putting on Christ: The *Toga Virilis* Ceremony, Its Paraenesis, and Paul's Interpretation of Baptism in Galatians." *Novum Testamentum* 44.3 (2002): 252–77.
Harvey, Susan Ashbrook. "Incense Offerings in the Syriac *Transitus Mariae:* Ritual and Knowledge in Ancient Christianity." In *The Early Church in its Context: Essays in Honor of Everett Ferguson*, eds. Abraham J. Malherbe, Frederick W. Norris, and James W. Thompson. Leiden: Brill, 1998.
———. "On Holy Stench: When the Odor of Sanctity Sickens." *Studia Patristica: Papers Presented to the International Conference on Patristic Studies* 35 (1998): 90–101.
———. *Scenting Salvation: Ancient Christianity and the Olfactory Imagination*. Berkeley: University of California Press, 2015.
Hauck, Robert. "They Saw What They Said They Saw: Sense Knowledge in Early Christian Polemic." *Harvard Theological Review* 81.3 (1988): 239–49.
Helms, Mary. "Ineffable Illumination: Early Medieval Church Treasure and the Preservation of Heaven's Light." *Anthropos* 109 (2014): 103–18.
———. "Sacred Landscape and the Early Medieval European Cloister: Unity, Paradise, and the Cosmic Mountain." *Anthropos* 97.2 (2002): 435–53.
Hen, Yitzhak. *Culture and Religion in Merovingian Gaul, 481–751*. New York: Brill, 1995.
Hendel, Ronald S. "Prophets, Priests, and the Efficacy of Ritual." In *Pomegranates and Golden Bells: Studies in Biblical, Jewish, and Near Eastern Ritual, Law and Literature in Honor of Jacob Milgrom*, eds. David P. Wright, David Noel Freedman, and Avi Hurvitz. Winona Lake, IN: J. E. Eisenbrauns, 1995, 185–98.
Hersch, Karen K. *The Roman Wedding: Ritual and Meaning in Antiquity*. Cambridge, UK: Cambridge University Press, 2010.
Heyman, George. *The Power of Sacrifice: Roman and Christian Discourses in Conflict*. Washington, DC: Catholic University of America Press, 2007.
Hidayatullah, Aysha. "Māriyya the Copt: Gender, Sex and Heritage in the Legacy of Muhammad's *Umm Walad*." *Islam and Christian-Muslim Relations* 21.3 (July 2010): 221–43.
Hoffman, Valerie J. "Muslim Sainthood, Women, and the Legend of Sayyida Nafisa." Ed. Arvind Sharma. *Women Saints in World Religions*. Albany: State University of New York Press, 2000, 107–44.
Hollender, Elisabeth. "Ritualization of Circumcision in Medieval Judaism in Relation to Islamic and Christianity: An Overview." *Religion* 42.2 (April 2012): 233–46.
Holmes, L. "Myrrh and Unguents in *Coma Berenices*." *Classical Philology* 87.1 (1992): 47–50.
Honarpisheh, Donna. "Women in Pilgrimage: Senses, Places, Embodiment, and Agency. Experiencing *Ziyarat* in Shiraz." *Journal of Shi`a Islamic Studies* VI.4 (2013): 383–410.

Hope, Valerie M. *Roman Death: The Dying and the Dead in Ancient Rome*. New York: Continuum, 2010.
Howes, David, and Constance Classen. *Ways of Sensing: Understanding the Senses in Society*. Toronto: Toronto University Press, 1991.
Houtman, C. "On the Function of the Holy Incense (Exodus XXX 34–8) and the Sacred Anointing Oil (Exodus XXX 22–33)." *Vetus Testamentum* XLII.4 (1992): 458–65.
Howes, David, ed. *Empire of the Senses: The Sensual Cultural Reader*. New York: Berg, 2005.
———. *The Varieties of Sensory Experience: A Sourcebook in the Anthropology of the Senses*. Toronto: University of Toronto Press, 1991.
Hunter-Crawley, Heather. "Embodying the Divine: The Sensational Experience of the Sixth-Century Eucharist." In *Making Sense of the Past: Toward a Sensory Archaeology*, ed. Jo Day. Carbondale: Southern Illinois University Press, 2013.
Hurowitz, Victor Avigdor. "Solomon's Golden Vessels (I Kings 7:48–50) and the Cult of the First Temple." In *Pomegranates and Golden Bells: Studies in Biblical, Jewish, and Near Eastern Ritual, Law, and Literature in Honor of Jacob Milgrom*, eds. David P. Wright, David Noel Freedman, and Avi Hurvitz. Winona Lake, IN: J. E. Eisenbrauns, 1995.
Jacob, Irene, and Walter Jacob, eds. *The Healing Past: Pharmaceuticals in the Biblical Rabbinic World*. New York: E. J. Brill, 1993.
Jacobs, Martin. "Römische Thermenkultur im Spiegel des Talmud Yerushalmi." In *The Talmud Yerushalmi and Graeco-Roman Culture*, v. 1, ed. Peter Schäferand Catherine Hezser. Tübingen: Mohr Siebeck, 1998.
———. "Theatres and Performances as Reflected in the Talmud Yerushalmi." In *The Talmud Yerushalmi and Graeco-Roman Culture*, v. 1, eds. Peter Schäfer and Catherine Hezser. Tübingen: Mohr Siebeck, 1998, 327–47.
James, Edward. "A Sense of Wonder: Gregory of Tours, Medicine and Science." In *The Culture of Christendom*, ed. Marc Anthony Meyer. Rio Grande: Hambledon Press, 1993, 45–60.
Jashemski, W. F. "The Garden of Hercules at Pompeii: The Discovery of a Commercial Flower Garden." *American Journal of Archaeology* 83: 403–11.
Jeanes, Gordon P. *The Day Has Come!: Easter and Baptism in Zeno of Verona*. Minnesota: Liturgical Press, 1995.
Jensen, Robin M. *Baptismal Imagery in Early Christianity: Ritual, Visual, and Theological Dimensions*. Grand Rapids, MI: Baker Academic Press, 2012.
Johanny, Richard. "Ignatius of Antioch." *Eucharist of the Early Christians*. New York: Pueblo Publishing Co., 1978.
Jordan, Mark D. *The Invention of Sodomy in Christian Theology*. Chicago: University of Chicago Press, 1997.
Jung, Dinah. *An Ethography of Fragrance: The Perfumery Arts of `Adan/Laḥj*. Boston: Brill, 2011.
Juynboll, G. H. A. "Dyeing the Hair and Bear in Early Islam: A Ḥadīth-Analytical Study." *Arabica*, Tome XXXIII (1986): 49–75.
———. *Studies on the Origins and Uses of Islamic Ḥadīth*. Hampshire: Variorum, 1996.
Katz, Marion Holmes. *Birth of the Prophet Muhammad: Devotional Piety in Sunni Islam*. Kentucky: Routledge, 2007.
———. *Body of Text: The Emergence of the Sunni Law of Ritual Purity*. Albany: State University of New York Press, 2002.
———. "The Study of Islamic Ritual and the Meaning of *Wuḍū'*." *Der Islam* 82 (2005): 106–45.
Keay, John. *The Spice Route: A History*. London: John Murray, 2005.
Kennedy, Hugh. "From Polis to Madina: Urban Change in Late Antique and Early Islamic Syria." *Past & Present* 6.106 (1985): 3–27.
King, Anya. "The Importance of Imported Aromatics in Arabic Culture: Illustrations from Pre-Islamic and Early Islamic Poetry." *Journal of Near Eastern Studies* 67.3 (2008): 175–89.
———. *The Musk Trade and the Near East in the Early Medieval Period*. Unpublished dissertation. Indiana University, 2007.

King, Charles. "The Organization of Roman Religious Beliefs." *Classical Antiquity* 22.2 (October 2003): 275–312.

Kleinschmidt, Harald. *Perception and Action in Medieval Europe*. Woolbridge: Boydill Press, 2005.

Kohl, James. *The Scent of Eros*. New York: Continuum, 1995.

Langer, Ruth. "The *Birkat Betulim:* A Study of the Jewish Celebration of Bridal Virginity." *Proceedings of the American Academy for Jewish Research* 61 (1995): 53–94.

Laurioux, Burno. "Cuisiner à L'Antique: Apicius au Moyen Âge." *Médiévales* 26 (1994): 17–38.

Lawrence, Louise. "Exploring the Sense-Scape of the Gospel of Mark." *Journal for the Study of the New Testament* 33.4 (2011): 387–97.

Leeper, Elizabeth A. "From Alexandria to Rome: The Valentinian Connection to the Incorporation of Exorcism as a Prebaptismal Rite. *Vigilae Christianae* 44 (1990): 6–22.

LeGuérer, Annick. "Le declin de l'olfactif: mythe ou réalité?" *Anthropologie det Sociétes* 14.2 (1990): 25–46.

———. *Scent: The Mysterious and Essential Powers of Smell*. New York: Kodansha International, 1994.

Lestocquoy, Jean. "Épices, medicine et abbayes." *Études mérovingiennes: Actes des journées de Poitiers, 1er–3 mai 1952*. Paris: A. et J. Picard, 1953.

Le Strange, Guy. *Palestine under the Moslems: A Description of Syria and the Holy Land*. Beirut: Khayat, 1965.

Levey, Martin. "Babylonian Chemistry: A Study of Arabic and Second Millenium B.C. Perfumery." *Osiris* 12 (1956): 376–89.

Levillain, Léon. *Examen critique des chartes mérovingiennes et carolingiennes de l'abbaye de Corbie* 15. Mémoires et documents publiés par la Société de l'École des chartes 5. Paris: A. Picard et Fils, 1902.

Lilja, Saara. *The Treatment of Odours in the Poetry of Late Antiquity*. Helsinki: Societas Scientiarum Fennica, 1972.

Luttikhuizen, Gerard P. ed. *Paradise Interpreted: Representations of Biblical Paradise in Judaism and Christianity*. Leiden: Brill, 1999.

Lybarger, Loren D. "The Demise of Adam in the *Qiṣaṣ al-Anbiyā':* The Symbolic Politics of Death and Re-Burial in the Islamic Stories of the Prophets." *Numen* 55 (2008): 497–535.

MacLean, Rachel, and Timothy Insoll. "Archaeology, Luxury, and the Exotic: The Examples of Islamic Gao (Mali) and Bahrain." *World Archaeology* 34 (3): 558–70.

Maghen, Ze'ev. "First Blood: Purity, Edibility, and the Independence of Islamic Jurisprudence." *Der Islam* 81.1 (2004): 49–95.

———. "Much Ado about *Wuḍū'*." *Der Islam* 79 (1999): 205–52.

———. *Virtues of the Flesh: Passion and Purity in Early Islamic Jurisprudence*. Boston: Brill, 2005.

Mahmoud, Ibrahim. *Jugrafiya al-maladhdhat al-jins bi al-janna*. Beirut: Riyad al-Rayyis. 1998.

de Maigret, A. "New Evidence from the Yemenite 'Turret Graves' for the Problem of Emergence of the South Arabian States." In *The Indian Ocean in Antiquity*, ed. J. E. Reade. London: Kegan Paul, 1996, 321–37.

Manson, T. W. "2 Cor. 14–17: Suggestions toward an Exegesis." In *Studia Paulina: In Honorem Johannis de Zwann Septuagenerii*, eds. J. N. Sevenster and W. C. van Unnik. Haarlem: de Ervem F. Bohn, 1953.

Mantle, I. C. "The Roles of Children in Roman Religion." *Greece & Rome*, Second Series, 49.1 (April 2002): 85–106.

Margoliouth, David. "The Relics of the Prophet Muhammad." *Moslem World* 27.1 (1937): 20–7.

Marín, Manuela. "Beyond Taste: The Complements of Colour and Smell in the Medieval Arab Culinary Tradition." In *Culinary Cultures of the Middle East*, eds. Sami Zubaida and Richard Tapper. New York: I. B. Tauris Publishers, 1994, 205–14.

Matin, Peyman. "Apotropaic Plants in the Persian Folk Culture." *Iran and the Caucasus* 16 (2012): 189–99.

Matthews, Thomas. *Clash of the Gods*. Princeton, NJ: Princeton University Press, 1993.
Mattingly, D. J. "Paintings, Presses, and Perfume Production at Pompeii." *Oxford Journal of Archaeology* 9: 71–90.
Maxwell-Stuart, P. G. "Myrtle and the Eleusinian Mysteries." *Wiener Studien* 85 (1972): 145–61.
McGowan, Andrew. "Discipline and Diet: Feeding the Martyrs in Roman Carthage." *Harvard Theological Review* 96.4 (2003): 455–76.
Memon, M. U. *Ibn Taymīya's Struggle against Popular Tradition; with an Annotated Translation of his Kitāb Iqtiḍā' al-Ṣirāṭ al-Mustaqīm Mukhālafat Ahl al-Jaḥīm*. Paris: Mouton & Co., 1976.
Meri, Josef W. *The Cult of Saints among Muslims and Jews in Medieval Syria*. Oxford: Oxford University Press, 2002.
———. "The Etiquette of Cult." In *The Cult of Saints in Late Antiquity and the Middle Ages*, eds. James Howard-Johnston and Paul Antony Hayward. New York: Oxford University Press, 1999, 263–86.
Meyerhof, Max. "'Ali aṭ-Ṭabarī's 'Paradise of Wisdom,' one of the oldest Arabic Compendiums of Medicine." *Isis* 16.1 (July 1931): 6–54.
Miller, J. Innes. *The Spice Trade of the Roman Empire, 29 BC to AD 641*. Oxford: Clarendon Press, 1969.
Miller, William Ian. *The Anatomy of Disgust*. Cambridge, MA: Harvard University Press, 1997.
Milwright, Marcus. "The Balsam of Maṭariyya: An Exploration of a Medieval Panacea." *Bulletin of the School of Oriental and African Studies* 66.2 (2003): 193–209.
Mintz, Sidney W., and Christine M. DuBois. "The Anthropology of Food and Eating." *Annual Review of Anthropology* 31 (2002): 99–119.
Moncrieff, R. W. *Odour Preferences*. New York: Wiley, 1966.
Motzki, Harald, ed. *The Biography of Muhammad: The Issue of the Sources*. Leiden: Brill, 2000.
Musallam, B. F. *Sex and Society in Islam: Birth Control before the Nineteenth Century*. Cambridge, UK: Cambridge University Press, 1983.
Neuwirth, A. "Face of God—Face of Man: The Significance of the Direction of Prayer in Islam." In *Self, Soul, and Body in Religious Experience*, eds. A. I. Baumgarten, J. Assmann, and G. G. Stroumsa. Leiden: Brill, 1998.
Newby, Gordon D. *"Tafsir Isra'iliyat:* The Development of Qur'an Commentary in Early Islam in Its Relationship to Judaeo-Christian Traditions of Scriptural Commentary." *Journal of the American Academy of Religion* XLVII.4 Thematic Issue S (December 1980): 685–97.
Nicholson, Oliver. "Preparation for Martyrdom in the Early Church." In *The Great Persecution: The Proceedings of the Fifth Patristic Conference, Maynooth, 2003*, eds. D. Vincent Twomey SVD and Mark Humphries. Dublin: Four Courts Press, 2009, 61–90.
Nir, Rivka. "The Aromatic Fragrances of Paradise in the Greek Life of Adam and Eve and the Christian Origin of the Composition." *Novum Testamentum* 40.1 (2004): 20–45.
Novick, Tzvi. "Peddling Scents: Merchandise and Meaning in 2 Cor. 14–17." *Journal of Biblical Literature* 130.3 (2011): 543–49.
Nutton, Vivian. *Ancient Medicine*. New York: Routledge, 2004.
Olson, Kelly. *Dress and the Roman Woman: Self-Presentation and Society*. New York: Routledge, 2008.
Olyan, Saul. "What Do Shaving Rites Accomplish and What Do They Signal in Biblical Ritual Contexts?" *Journal of Biblical Literature* 117.4 (1998): 611–22.
Orlov, Andrei A. *From Apocalypticism to Merkabah Mysticism: Studies in the Slavonic Pseudepigrapha*. Leiden: Brill, 2007.
Pagan, Victoria Emma. *Rome and the Literature of Gardens*. London: Duckworth, 2006.
Parker, W. H. *Priapea: Poems for a Phallic God*. London: Croom Helm Ltd., 1988.
Peacock, David, and David Williams. *Food for the Gods: New Light on the Ancient Incense Trade*. Oxford: Oxbow Books, 2007.
Pettis, Jeffrey B. "Earth, Dream, and Healing: The Integration of *Materia* and *Psyche* in the Ancient World." *Journal of Religion and Health* 45.1 (Spring 2006): 113–29.

Pigeon. M. E. A. "De l'embaumement des morts à l'époque mérovingienne." *Bulletin archéologuique du Comité des travaux historiques et scientifiques* (1894): 139–42.
Regev, Eyal. "Archaeology and the Mishnah's Halakhic Tradition: The Case of Stone Vessels and Ritual Baths." In *The Mishnah in Contemporary Perspective, V. 2*, eds. Alan J. Avery-Peck and Jacob Neusner. Boston: Brill, 2006, 136–52.
Reich, Ronny. "The Synagogue and the *Miqweh* in Eretz-Israel in the Second-Temple, Mishnaic, and Talmudic Periods." In *Ancient Synagogues: Historical Analysis and Archaeological Discovery, V. 1*, eds. Dan Urman and Paul V. M. Flesher. New York: Brill, 1995, 289–97.
Reinhart, Kevin. "Impurity/No Danger." *History of Religions* 30 (1990): 1–24.
Richardson, Kristina. "Singing Slave Girls *(Qiyan)* of the `Abbasid Court in the Ninth and Tenth Centuries." In *Children in Slavery through the Ages*, eds. Gwyn Campbell, Suzanne Miers, and Joseph C. Miller. Athens: Ohio University Press, 2009.
Riddle, John. *Contraception and Abortion from the Ancient World to the Renaissance*. Boston: Harvard University Press, 1992.
Rindisbacher, Hans J. *The Smell of Books: A Cultural-Historical Study of Olfactory Perception in Literature*. Ann Arbor: University of Michigan Press, 1992.
Robinson, Chase. "Prophecy and Holy Men in Early Islam." In *The Cult of the Saints in Late Antiquity and the Middle Ages*, eds. James Howard-Johnston and Paul Antony Hayward. Oxford: Oxford University Press, 1999.
Rosenblum, Jordan D. *Food and Identity in Early Rabbinic Judaism*. Cambridge, UK: Cambridge University Press, 2010.
Rosivach, Vincent J. "Lex Fannia Sumptuaria of 161 BC." *The Classical Journal* 102.1 (2006): 1–15.
Rossetti, Chip. "Devil's Dung: The World's Smelliest Spice." *Saudi Aramco World*, July/August (2009): 36–43.
Rothkrug, Lionel. *Death, Trust, & Society: Mapping Religion & Culture*. Berkeley: North Atlantic Books, 2006.
———. "The Odour of Sanctity and the Hebrew Origins of Christian Relic Veneration." *Historical Reflections* 8.2 (1981): 95–142.
Rouselle, Aline. *Porneia: On Desire and the Body in Antiquity*. Cambridge, UK: Blackwell Publishers, 1993.
Rowson, Everett K. "Gender Irregularity as Entertainment: Institutionalized Transvestism at the Caliphal Court in Medieval Baghdad." *Gender and Difference in the Middle Ages*. Minneapolis: University of Minnesota Press, 2003.
———. "The Effeminates of Early Medina." *Journal of the American Oriental Society* 111.4 (1991): 671–93.
Rubin, Nissan. *Time and Life Cycle in Talmud and Midrash: Socio-Anthropological Perspectives*. Boston: Academic Studies Press, 2008.
Rubin, Uri. "Pre-existence and Light: Aspects of the Concept of the Nūr Muḥammad." *Israel Oriental Studies* 5 (1975): 62–119.
Rudy, Gordon. *Mystical Language of the Senses in the Later Middle Ages*. New York: Routledge, 2002.
Ruggles, Fairchild. *Gardens, Landscape, and Vision in the Palaces of Islamic Spain*. Pittsburgh: University of Pennsylvania Press, 2011.
———. *Islamic Gardens and Landscapes*. Philadelphia: University of Pennsylvania Press, 2011.
Rustomji, Nerina. *The Garden and the Fire: Heaven and Hell in Islamic Culture*. New York: Columbia University Press, 2012.
Safrai, Zeeve. "The Communal Functions of the Synagogue in the Land of Israel in the Rabbinic Period." In *Ancient Synagogues: Historical Analysis and Archaeological Discovery V. 1*, eds. Dan Urman and Paul V. M. Flesher. New York: Brill, 1995, 181–204.
Saiko, M. "Cura dabit faciem: Kosmetik in Altertum-literarishe, kulturhistorische und medizinische Aspekte." *Bochumer Altertumswissenschaftliches Colloquium 66*, Trier, 2005.
Sanders, Paula. *Ritual, Politics, and the City in Fatimid Cairo*. Albany: State University of New York Press, 1994.

Satlow, Michael L. "'They Abused Him Like a Woman': Homoeroticism, Gender Blurring, and the Rabbis in Late Antiquity." *Journal of the History of Sexuality* 5.1 (July 1994): 1–25.
Schab, Frank R., and Robert G. Crowder. *Memory for Odors*. Mahwah, NJ: Lawrence Erlbaum Associates, 1995.
Schimmel, Annemarie. *And Muhammad Is His Messenger*. Chapel Hill: University of North Carolina Press, 1985.
Schwiebert, Jonathan. *Knowledge and the Coming Kingdom: The Didache's Meal Ritual and its Place in Early Christianity*. New York: T&T Clark, 2008.
Seeley, David. *The Noble Death: Graeco-Roman Martyrology and Paul's Concept of Salvation*. Sheffield: Sheffield Academic Press, 1990.
Seiler, Thomas. "Filth and Stench as Aspects of the Iconography of Hell." In *The Iconography of Hell*, eds. Clifford Davidson and Thomas H. Seiler. Kalamazoo: Medieval Institute Publications, 1992.
Sells, Michael. *Desert Tracings: Six Classic Arabian Odes*. Connecticut: Wesleyan University Press, 1986.
———. "Guises of the Ghūl." In *Reorientations/Arabic and Persian Poetry*, ed. Suzanne Pinckney Stetkevych. Bloomington: Indiana University Press, 1994.
Shanzer, Danuta. "Bishops, Letters, Fast, Food, and Feast in Later Roman Gaul." In *Society and Culture in Late Antique Gaul: Revisiting the Sources*, eds. Ralph W. Mathisen and Danuta Shanzer. Burlington: Ashgate, 2001, 217–36.
Shepkaru, Shmuel. "From after Death to Afterlife: Martyrdom and Its Recompense." *AJS Review* 24.1 (1999): 1–44.
Sindawi, Khalid. "The Head of Husayn ibn `Ali: Its Various Places of Burial and the Miracles that it Performed." In *Holy Places in the Israeli-Palestinian Conflict*, eds. Marshall J. Breger, Yitzhak Reiter, and Leonard Hammer. New York: Routledge, 2010.
Smith, Dennis. *From Symposium to Eucharist: The Banquet in the Early Christian World*. Minneapolis: Fortress Press, 2003.
Smith, Jane, and Yvonne Haddad. "Women in the Afterlife: The Islamic View as Seen from Qur'an and Tradition." *Journal of the American Academy of Religion* 43.1 (1975): 39–50.
Sorensen, Eric. *Possession and Exorcism in the New Testament and Early Christianity*. Mohr Siebeck: Tübingen, 2002.
Stetkevych, Jaroslav. *The Zephyrs of Najd: The Poetics of Nostalgia in the Classical Arabic Nasīb*. Chicago: University of Chicago Press, 1993.
Stetkevych, Suzanne Pinckney. *The Mute Immortals Speak: Pre-Islamic Poetry and the Poetics of Ritual*. Ithaca, NY: Cornell University Press, 1993.
———. *The Poetics of Islamic Legitimacy: Myth Gender and Ceremony in the Classical Arabic Ode*. Bloomington: Indiana University Press, 2002.
———. *Reorientations/Arabic and Persian Poetry*. Ed. Suzanne Pinckney Stetkevych. Bloomington: Indiana University Press, 1994.
Steuer, Robert O. "Stactē in Egyptian Antiquity." *Journal of the American Oriental Society* 1.4 (October–December 1943): 279–84.
Stoller, Paul. *The Taste of Ethnographic Things*. Philadelphia: University of Pennsylvania Press, 1989.
Stowasser, Barbara. "The Mothers of the Believers in the Ḥadīth." *Muslim World* 82.1–2 (1992): 1–36.
Synnott, Anthony. *The Body Social: Symbolism, Self, and Society*. New York: Routledge, 1993.
Tabor, James D. *Things Unutterable: Paul's Ascent to Paradise in Its Greco-Roman, Judaic, and Early Christian Contexts*. New York: University Press of America, 1986.
Taylor, Christopher S. *In the Vicinity of the Righteous: Ziyāra and the Veneration of Muslim Saints in Late Medieval Egypt*. Leiden: Brill, 1999.
Thompson, Leonard L. "The Martyrdom of Polycarp: Death in the Roman Empire." *Journal of Religion* 82.1 (2002): 27–52.
Thurlkill, Mary. *Chosen among Women: Mary and Fatima in Medieval Christianity and Shi`ite Islam*. Notre Dame, IN: Notre Dame Press, 2007.
Tigchelaar, J. C. "Eden and Paradise: The Garden Motif in Some Early Jewish Texts (I Enoch and Other Texts Found at Qumran)." In *Paradise Interpreted: Paradise Interpreted: Repre-*

sentations of Biblical Paradise in Judaism and Christianity, ed. Gerard P. Luttikhuizen. Leiden: Brill, 1999.

Toner, Jerry. *Popular Culture in Ancient Rome*. Malden, MA: Polity Press, 2009.

Tottoli, Robert. "Tours of Hell and Punishments of Sinners in *Mi`raj* Narratives: Use and Meaning of Eschatology in Muhammad's Ascension." In *The Prophet's Ascension: Cross-Cultural Encounters with the Islamic Mi`raj Tales*. Bloomington: Indiana University Press, 2010.

Toynbee, J. M. C. *Death and Burial in the Roman World*. London: Thames and Hudson, 1971.

Treggiari, Susan. *Roman Marriage: Iusti Coniuges from the Time of Cicero to the Time of Ulpian*. Oxford: Clarendon Press, 1991.

Turner, Jack. *Spice: The History of a Temptation*. New York: Alfred A. Knopf, 2004.

Turner, Victor. *The Ritual Process: Structure and Anti-Structure*. Chicago: Aldine Transactions, 1969.

Twelftree, Graham H. *In the Name of Jesus: Exorcism among Early Christians*. Grand Rapids, MI: Baker Academic, 2007.

Tyrrell, W. Blake. *The Smell of Sweat*. Mundelein, IL: Bolchazy-Carducci Publications, Inc., 2004.

Van Gelder, G. J. H. "Four Perfumes of Arabia." In *Parfums d'Orient. Textes reunis par R. Gyselen*. Bures-sur-Yvette: Groupe pour l'étude de la civilization du moyen-orient, 1998.

Van Slyke, Daniel G. "The Devil and His Pomps in Fifth-Century Carthage: Renouncing *Spectacula* with Spectacular Imagery." *Dumbarton Oaks Papers* 59 (2005): 53–72.

Vogel. "Deaux consequences de l'eschatologie grégorienne: La multiplication des messes privies et les moines-prêtres." In *Grégoire le Grand: Chantilly, Centre culturel les Fontaines*, eds. Jacques Fontaine, Robert Gillet, Stan Pellistrandi, Colloques internationaux du CNRS. September 15–19, 1982. Paris: Editions du CNRS, 1986.

Vogt, Burkhard. "Death and Funerary Practices." In *Caravan Kingdoms: Yemen and the Ancient Incense Trade*, ed. Ann C. Gunter. Washington, DC: Smithsonian Institution, 2005, 81–88.

Vons, J. *L'image de la femme dans l'oeuvre de Pline L'Ancien*. Brussels: Collection Latomus 256, 2000.

Vos, Nienke, and Willemien Otten, eds. *Demons and the Devil in Ancient and Medieval Christianity*. Boston: Brill, 2011.

Waines, David. "Cereals, Bread and Society: An Essay on the Staff of Life in Medieval Iraq." *Journal of the Economic and Social History of the Orients* 30.3 (1987): 255–85.

———. "'Luxury Foods' in Medieval Islamic Societies." *World Archaeology* 34.3 (2003): 571–80.

Ward, Roy Bowen. "Women in Roman Baths." *Harvard Theological Review* 85.2 (1992): 125–47.

Al-Washshā, Ahmad. *Kitāb al-Muwashshā*. Ed. R. E. Brünnow. Leiden: Brill, 1886.

Watson, Andrew M. "The Arab Agricultural Revolution and Its Diffusion, 700–1100." *Journal of Economic History* 34.1 (1974): 8–35.

Wheeler, Brannon. "Arab Prophets of the Qur'an and Bible." *Journal of Qur'anic Studies* 8.2 (2006): 24–57.

———. "Gift of the Body in Islam: The Prophet Muhammad's Camel Sacrifice and Distribution of Hair and Nails at His Farewell Pilgrimage." *Numen* 57 (2010): 341–88.

———. *Mecca and Eden: Ritual, Relics, and Territory in Islam*. Chicago: Chicago University Press, 2006.

———. "Touching the Penis in Islamic Law." *History of Religions* 44 (2004): 89–114.

Wilson, Donald. *Learning to Smell*. Baltimore: John Hopkins University, 2006.

Yasin, Ann Marie. *Saints and Church Spaces in the Late Antique Mediterranean: Architecture, Cult, and Community*. Cambridge, UK: Cambridge University Press, 2009.

Young, Robin Darling. *In Procession before the World: Martyrdom as Public Liturgy in Early Christianity*. Milwaukee: Marquette University Press, 2001.

Index

About the Author

Mary Thurlkill is associate professor of Religious Studies at the University of Mississippi. She earned her PhD in Religion from Indiana University and authored *Chosen among Women: Mary and Fatima in Medieval Christianity and Shi`ite Islam* (2007).

www.ingramcontent.com/pod-product-compliance
Lightning Source LLC
Chambersburg PA
CBHW051252200926
51612CB00001B/18
9780739174524